SOTHIS PRESS

Sothis Press aims to completely reimagine the public domain book with an emphasis on universal access to captivatingly designed print editions. We believe that technology will enable every single book ever published to be perpetually accessible in beautiful print form. Sothis is beauty in ubiquity.

www.sothis.press
1 Waterhouse Square, London, ECIN 2ST

Accurate Tool Work

C. L. Goodrich

SOTHIS PRESS

ISBN 978-1-0368-0096-3

This edition is published by
Sothis Press © 2023. The company
is registered in England and Wales.
The Company Number is 14900389.

ACCURATE TOOL WORK

BY

C. L. GOODRICH

Department Foreman, Pratt and Whitney Company

AND

F. A. STANLEY

Associate Editor American Machinist; Author of "*The Hill Kink Books,*" and "*American Machinists' Handbook*"

Published by the

McGraw-Hill Book Company

New York

Successors to the Book Departments of the

McGraw Publishing Company   Hill Publishing Company

Publishers of Books for

Electrical World   The Engineering and Mining Journal

The Engineering Record   Power and The Engineer

Electric Railway Journal   American Machinist

The Hill Publishing Company, New York, U.S.A.

PREFACE

THE occupation of the toolmaker has steadily grown in importance with the development of manufacturing processes and the general adoption of jigs, fixtures and other special tools by machinery builders to insure interchangeability of product. Where formerly but few men in the shops were directly interested in, for example, the methods by which the holes in a jig could be accurately located and bored, to-day there are thousands of toolmakers who are employing refined processes, precision tools and appliances for executing this class of work. Many methods and devices originating in watch factories and similar establishments for accomplishing very accurate results were for a considerable period confined almost exclusively to such institutions; these have in later years been found equally serviceable in dealing with work of different character and heavier proportions.

The master plate, disk, button and refined test indicator processes have been extended from watch-tool to other classes of accurate tool work, and these invaluable adjuncts of the tool room have before them an ever broadening field of usefulness as their practicability becomes more generally appreciated. Closely allied with these devices, although up to the present time utilized to a comparatively small extent, is the compound microscope, which with cross hairs and conveniently arranged micrometer screws constitutes a testing and measuring appliance having an innumerable number of practical applications in connection with the work of the toolmaker.

We have endeavored in the following pages to present, in convenient form, information on various phases of tool work contained in articles published in the past few months in the *American Machinist* and prepared originally by different contributors and members of the editorial staff of that journal. Recognizing the impossibility of covering even superficially in a volume of this character the wide variety of operations falling naturally within the province of the toolmaker, we have confined the subject-matter of this book to the field suggested in a general way in the preceding paragraphs.

There is no branch of tool work more important or more interesting than that relating to the construction of jigs and other special tools, in which some of the most ingenious and precise methods known to the skilled mechanic are used. These methods, including the use of master plates, buttons, disks, size blocks, etc., have therefore been treated at length, together with processes of making master plates for various purposes, the use of test indicators, accurate gages, the microscope and other appliances, and considerable space has been devoted to ways and means of dealing with angular and taper work. The sections pertaining to the latter subjects contain information which should be of service in connection with the computing of angles, the finding of distances between hole centers, the lengths of sloping sides and other dimensions frequently required in tool work. The applications of disks and plugs in the measurement of tapers and dovetails are, owing to their importance in this work, also illustrated in detail.

It is really surprising that, considering the great development to which these methods have been carried, permanent literature relating to them scarcely exists. Much still remains to be said, but we have at least endeavored to place before the reader information of importance to every toolmaker and heretofore not accessible except through personal contact with work of the nature described.

We acknowledge our indebtedness to the authors of various articles incorporated with our own in this book, and in foot notes to various chapters individual credit is given.

THE AUTHORS.

CONTENTS

CHAPTER I

Locating and Boring Holes in Drill Jigs

There are various methods in common use by toolmakers for locating and boring holes in drill jigs, the particular method selected depending upon the size of the jig, the degree of accuracy required, and frequently upon the tool equipment available.

Among the methods of doing this work may be mentioned:

1. Laying out by scale, dividers and center punch. (Usually boring in drill press or lathe.)
2. Laying out by scriber and hight gages. (Usually boring in drill press or lathe.)
3. Size-block method. (Lathe or milling machine.)
4. Button method. (Lathe or milling machine.)
5. Screw and micrometer-dial method. (Milling machine.)
6. Micrometer depth-gage method. (Milling machine.)
7. Vernier caliper method. (Milling machine.)
8. Vernier hight-gage method. (Milling machine.)
9. Disk method. (Lathe.)
10. Transfer or master-plate method. (Lathe, generally.)

There are also a number of special methods, not in as common use as the foregoing, which will be touched upon.

LAYING OUT BY SCALE, DIVIDERS AND CENTER PUNCH

Very little need be said about the use of these tools on jig work, although in the hands of a skilful workman very close measurements may be obtained. We know of at least one toolmaker who, in setting distances by dividers, always when possible scribes a full circle from the center-punch mark and measures across the full diameter, believing he can work much closer by doubling up in this manner. In center-punching holes it is advisable to use a cylindrical punch, as it may be held more squarely.

THE SCRIBER BLOCK AND VERNIER HIGHT GAGE

Previous to prick punching the centers of the various holes that are required to be bored in an accurate jig, center lines are scratched by means of a scriber attached to a block, and various hights or measurements obtained by resting the base of the scribe block on size blocks varying in thickness to suit the various center distances; the jig plate during this operation rests on a surface plate.

The Brown & Sharpe vernier hight gage with scriber attached is a most convenient adjustable tool for accurately scribing lines prior to prick punching. A careful toolmaker in using the punch first puts an exceedingly small center at the intersection of the lines, then verifies the accuracy of its position by aid of a glass, and if correct then sinks the punch in deeper, holding the punch at this time at right angles to the jig plate. A V-square, such as is illustrated in Fig. 1, is used in some shops and is quite convenient for guiding the punch.

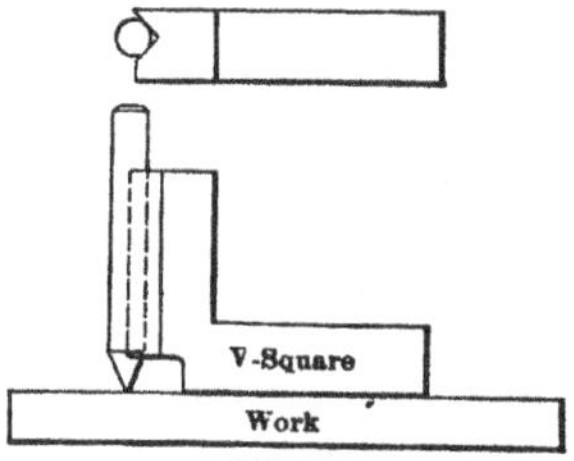

FIG. 1

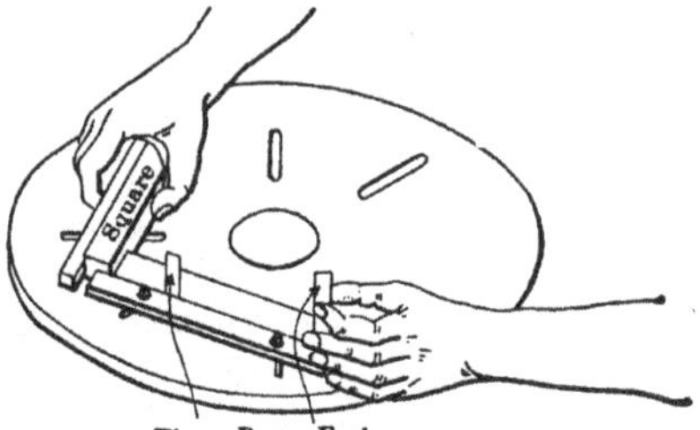

FIG. 5

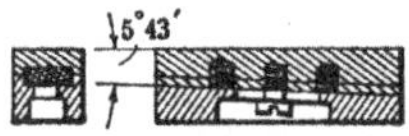

FIG. 4

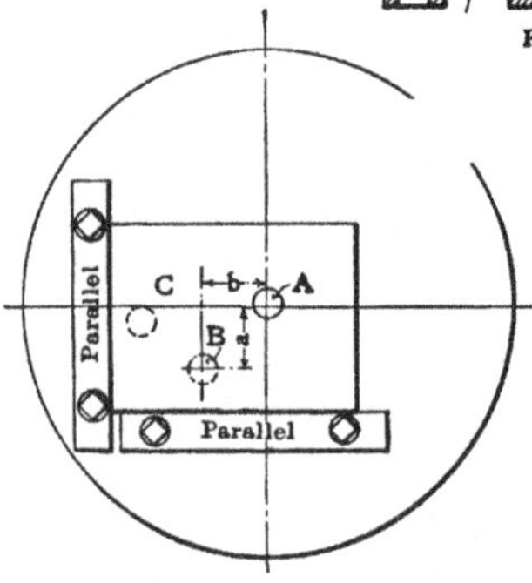

FIG. 2

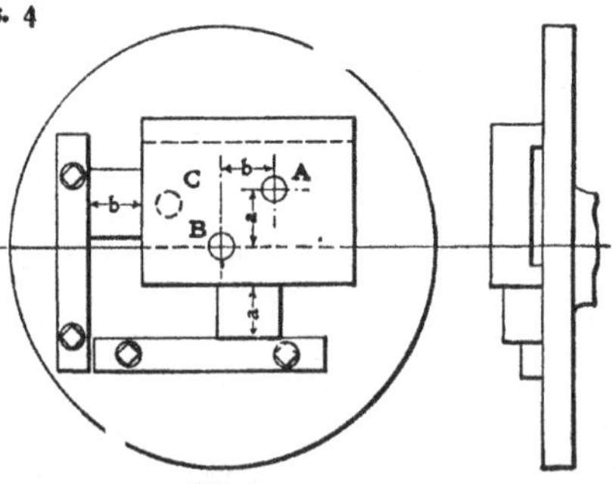

FIG. 3

LOCATING AND BORING HOLES IN DRILL JIGS

In case the bush holes have been accurately centered by the punch and the drilling of the holes is to be done in a drill press, it is conducive to accuracy first to drill the various holes with a small drill, say $\frac{1}{32}$ to $\frac{3}{32}$ inch diameter (or slightly less in diameter than the width of web on the reamer drill) prior to the use of the reamer drill, as the latter, on account of the center web, will not follow the punch center as closely as under the two-drill method. Most toolmakers scribe a circle representing the outside diameter of each hole before drilling out the center-punch mark with the small drill, the circle assisting them in watching the travel of the larger drill.

It is essential, in case a drill press is used to drill and ream the holes in a jig, that the table of the press be trammed up square with the spindle; if this is done and the spindle is closely fitted, jigs may be laid out, drilled and reamed within 0.004 to 0.005 inch of correct center distances, this depending upon the size of holes and material the jig is made of.

In case the jig is strapped on the face plate of a lathe and carefully "indicated up," somewhat closer work will be assured. For a large percentage of work the center-punch method will be found entirely satisfactory.

THE SIZE-BLOCK METHOD

This method, where a good set of size blocks is at hand, is found to be one of the most rapid and accurate.

The jig proper, as in all cases where adjustments or measurements are to be taken from the edges, should be carefully planed square and parallel. Considering now that the work is to be done on a lathe, it is essential that the face plate be carefully trued up and that it should fit the spindle tightly with absolutely no shake. The spindle of the lathe obviously should be a close running fit in the boxes, as with all work requiring accuracy.

A couple of good parallels should be located on the face plate at right angles to each other, as shown in Fig. 2, their position being such as to locate the jig so that the hole nearest the upper and outer edges of the jig (hole *A* in Fig. 2) is in line with the axis of the head spindle. This hole may then be bored.

Subsequent locations of the jig are accomplished by interposing size blocks between the two edges of the parallels and the jig, as shown in Fig. 3, the thickness of the size blocks as indicated by dimensions *a* and *b* being equal to the two dimensions *a* and *b* between the center lines of holes *A* and *B*.

The jig is located for boring hole *C* by using size blocks whose thicknesses correspond to the center distances between holes *A* and *C*. No matter how many holes are required in the jig, they may all be readily bored by this method, locating the work entirely from the two edges against the parallels on size blocks. For some work, where there are a good many center distances varying by thousandths, adjustable blocks like Fig. 4 may be of service. This style of block is easily set to thickness by sliding one member on the other and measuring the thickness by the micrometer, the screw being set up lightly to hold the two members when adjusted properly. If the tongue is made separately and inserted, the two sloping surfaces can be lapped more readily. An angle of 5 degrees, 43 minutes, is a convenient one for the contacting surfaces, for with this angle a longitudinal adjustment of 1 inch means an increase or decrease in hight or thickness of 0.1 inch and so on. Gages of the type shown in Chapter XVIII should also be of value in setting the work at the required distances from the fixed parallels, as these gages may be combined and used as size blocks, giving by ten-thousandths any thickness desired.

Oftentimes where size blocks are available that are nearly but not quite to the required thickness, paper of the necessary thickness is used to make up the difference. It is the practice of some toolmakers to use a piece of tissue paper between the parallel and size blocks in locating the jig prior and subsequent to strapping the work to the face plate; as a very slight change in position is detected by the loosening or tightening of the paper, it is recommended for very close work. The placing of the parallels exactly at right angles to each other is readily accomplished by the use of a perfect square and tissue-paper feelers, as indicated by Fig. 5.

Where accurate blocks are used and a careful toolmaker does the planing on the jigs and the boring of the holes, the method outlined may be depended upon to locate holes within 0.0005 inch on jigs up to say 8x10 inches. Where the jigs are much larger than these figures, the use of the milling machine in conjunction with size blocks or plugs is found to produce the better results.

LARGER WORK ON THE MILLER

Suppose we have a jig plate or a plate jig of too large dimensions to swing on the lathe and decide to use the size-block method or a similar scheme for locating it correctly on the miller while its various holes are being bored. An accurate way to handle the job will be to mount it on a sub-plate which can in turn be secured to the vertical face of a knee or angle plate which is bolted to the miller table, both the table and the miller knee being locked fast. All adjustments of the work from hole to hole are accomplished by shifting the sub-plate on the face of the knee, the various positions of the sub-plate being gotten by the use of either size blocks or end measure gages.

Referring to Fig. 6, *A* is the knee tongued and secured to the miller table *B*, and *C* is the sub-plate carrying jig *D*. The cords attached to the sub-plate are connected with a weight by which the plate and work are counterbalanced so that they may be easily shifted about on the face of the knee. Steel strips *E* and *F* are secured to knee *A*, and similar strips *G* and *H* to sub-plate *C*. These strips are finished with great care to serve as permanent straight-edges, their working edges being scraped perfectly straight and true with one another. The vertical edges are cut down to form guide surface, as indicated by *a*, *b* and *c*, and the horizontal

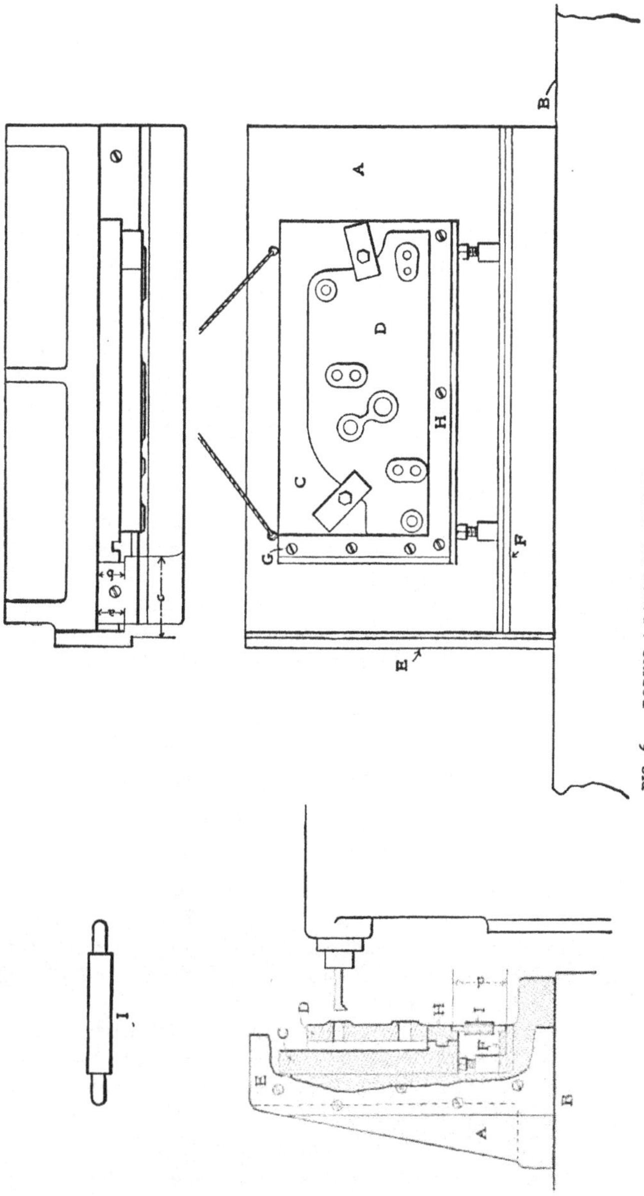

FIG. 6. — BORING A LARGE JIG ON THE MILLER

strips are finished in similar fashion. Between the shoulders thus formed on the pairs of strips, the gages are placed when setting the work for boring. As distances *a* and *b* are made exactly alike, the guides thus formed by these ledges enable the gages to be placed parallel with the face of the knee and sub-plate, insuring the latter being adjusted the exact horizontal distance required for each successive setting. The guiding surfaces of the straight-edges *F* and *H* insure correct setting vertically by the distance gages or plugs.

A convenient form of plug is shown at *I*, the ends being turned smaller than the body, finished spherically and lapped to the exact length required. With the hole at the end of the jig taken as the starting point, the gages may be made to suit the different horizontal and vertical distances; as the working edges of the straight-edges are corrected when first applied, until they are as nearly perfect as possible, very accurate results can be obtained by this method. The vertical adjustment of the work is facilitated by small jacks and the counterweight already referred to. The work is of course secured by clamping the sub-plate very carefully to the knee after each setting.

Size blocks may be used in various cases for obtaining the required distances from hole to hole, and in some work inside micrometers and other forms of internal measuring appliances will prove convenient. It is apparent that for many jobs the extra refinement made possible by the corrected guide surfaces on the straight-edges can be dispensed with and sufficiently accurate results obtained by parallels which are secured to sub-plate and knee and set square with each other by the method illustrated in Fig. 5.

It may be stated here that both the vertical-spindle milling machine and the vertical-spindle attachment for the universal miller are ideal for accurately boring holes in jigs.

THE BUTTON SCHEME

Probably the most common method used at present to locate jigs on the face plate of a lathe is the button scheme, which involves the accurate positioning and clamping of cylindrical bushings to the face of the jig by means of small screws, as indicated by Fig. 7, and utilizing the outer diameter of the bushing for an indicator to ride on while locating the jig on the face plate. It is obvious that if the bushings are so located that their axes are continuations of the axes of the holes that are to be bored, and providing that the jig be positioned previously to boring each hole so that an indicator resting on the outside of a bushing shows that the latter runs perfectly true, each hole will be bored in the proper location.

The buttons or bushings should be ground so that the outer surface is perfectly square with the ends. A common size for buttons is, outside diameter $\frac{1}{2}$ inch, diameter of hole $\frac{5}{16}$ inch, thickness $\frac{3}{8}$ inch. The clamping-screw dimensions recommended in connection with the foregoing buttons are $\frac{3}{16}$ body, $\frac{7}{16}$ head, $\frac{3}{4}$ inch long under the head.

After the jig has been carefully planed, the toolmaker lays out the centers of all holes by scale and dividers within approximately 0.010 inch, and drills and taps the button clamping-screw holes. Next, after carefully removing all burs and dust, he places a button on the face of the jig and lightly clamps it with the screw. Then by micrometers, or verniers, and size blocks he adjusts the button until its axis is known to be located just central for the hole that is to be bored. The clamping screw is then more tightly screwed home, and, providing that subsequent measurements show no change of position of the button, the jig may be clamped to the face plate after being carefully positioned by means of an indicator running on the outer diameter of the button, as in Fig. 8. The remaining holes are put in similarly. It may be well to point out that there is less danger of accumulated errors by taking all measurements from two points on the jig, as for instance from parallels against the sides of the jig, as shown in Fig. 9.

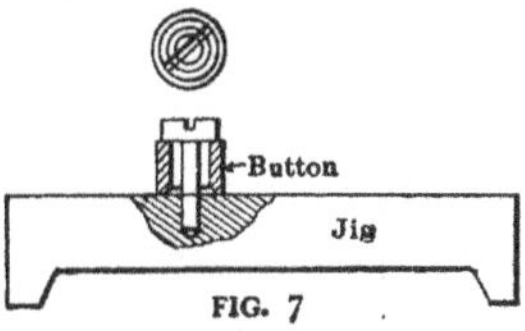

FIG. 7

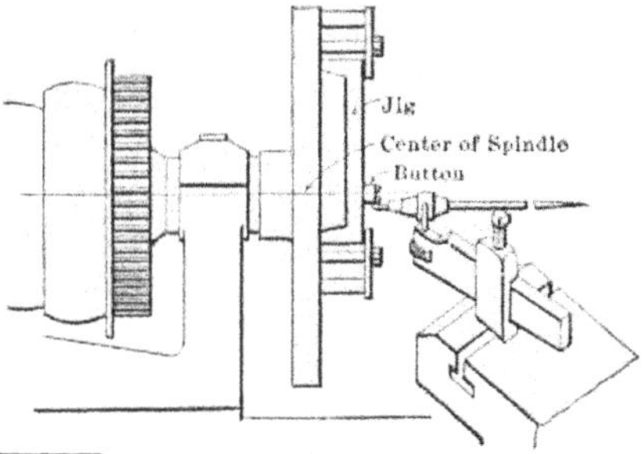

FIG. 8

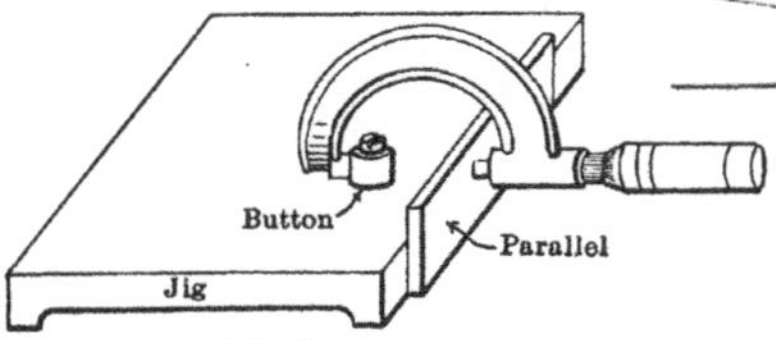

FIG. 9

THE BUTTON METHOD ON JIG WORK

Where measurements are taken from the nearest hole a number of minute errors frequently result. In all accurate face-plate work it is assumed that the weight of the jig is well balanced by a counterweight; otherwise the centrifugal action, in the case of slight looseness in the spindle, may introduce errors when the lathe is speeded up.

Jigs bored in the foregoing manner can be depended upon to show not over 0.0005 inch error in the center distances.

For boring holes in a jig on a lathe, the button scheme is recommended in preference to the size-block method when the shape of the jig is very irregular and the work projects quite a distance from the face toward the tailstock. Even when there is very little play to the spindle and the face plate is well balanced, the centrifugal action is sufficient to introduce errors when boring a jig positioned by a size block; if the button scheme is utilized and the jig is indicated with the lathe running at boring speed, the error is minimized, although not eliminated, as the vibrating action of the jig causes the axis of the hole, however bored, to be out. A jig whose shape is such as to cause springing action when revolving should, if possible, be bored on the boring mill or milling machine with a rotating cutter, the jig itself being stationary.

The button scheme is sometimes used in connection with a milling machine on larger jigs than can be swung in the engine lathe, as shown by Fig. 10. Referring to the sketch, A is a stiff and true running arbor,

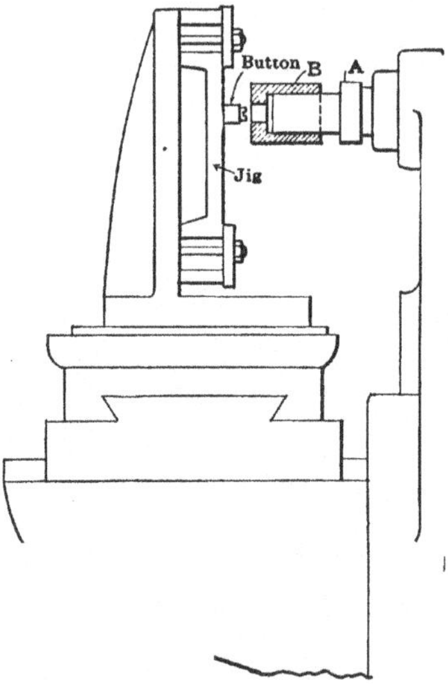

FIG. 10. — THE BUTTON SCHEME ON THE MILLER

B is a sliding sleeve with a hole ground exactly 0.00025 inch larger than the outside diameter of the button. The jig is located by means of the sliding bushing, which is pushed over the various buttons on the jig as each hole is to be bored. If the arbor is put into the spindle in the same position each time and the spindle is stopped at exactly the same point each time the jig is to be set, and if the toolmaker uses a reasonable amount of care, no difficulty will be experienced in boring holes to within 0.00075 inch of the desired position. Much depends upon the sensitiveness of the workman's touch. The knee to which the jig is strapped while being bored must be trammed square with the spindle, and the saddle should also travel parallel with the center of the spindle in order to produce good results.

There have been devised special types of indicators for use in the miller spindle for setting jigs by means of the buttons. A common difficulty with such indicators is that the reading scale revolves as the spindle is turned, and it is difficult to note the amount of fluctuation of the pointer during the complete rotation of the instrument. On certain classes of work, however, such indicators may prove of considerable service.

USING THE MILLING-MACHINE SCREW

In some shops practically all jigs are bored on the universal milling machine, the locating of the jig being accomplished by means of the knee-elevating and table-feed screws. Where these screws are carefully made and taken care of, good results are obtained, the graduated dials giving conveniently the required settings from center to center of the holes to be bored.

Where the miller screws are not sufficiently accurate for the work in hand, a method may be adopted which produces accuracy irrespective of the condition of the screws themselves, although the adjustments are made by these screws and their dials. If two holes, say, are to be bored a certain distance apart, the first may be bored with the dial of the table screw set at zero and the table then moved until according to the dial the work is at the required point for boring the second hole. This hole is then bored small, a plug inserted in it and another plug placed in the first hole. A measurement is then made by micrometer or vernier over the outside of the two plugs; by subtracting half the diameter of each plug from the reading of the measuring instrument, the exact center distances to which the holes have been bored will be readily obtained. Now suppose that, because of inaccuracy in the screw, the distance as measured is found to be a little longer or shorter than the center distance actually required. The screw is turned to readjust the table in the right direction the required amount as shown by the dial, and the hole bored out to size. In particularly accurate work the first hole is sometimes bored small at the outset and the second hole bored to the same size, so that duplicate plugs are used in making the trial measurement. After readjustment of the table (if this is necessary) the second hole is then finished to size and the table run back a little past the original setting point where the first hole was bored. The screw is then turned ahead, taking out the back lash and bringing the dial again to zero, which sets the work again in its original position. The first hole is then bored to size, completing the job. With this process, as with other jig-boring methods on the milling machine where the table and knee are adjusted to obtain the required center distances for the holes, the gibs must be adjusted nicely to give the bearing surfaces a snug even movement and yet insure the table retaining its position when set without the clamping screws being locked tight.

USE OF MICROMETER DEPTH GAGES

Two 1-inch micrometer depth gages are sometimes fastened firmly and at right angles to each other on a stationary angle plate on the milling machine, as shown in Fig. 11. By means of these micrometers and a set

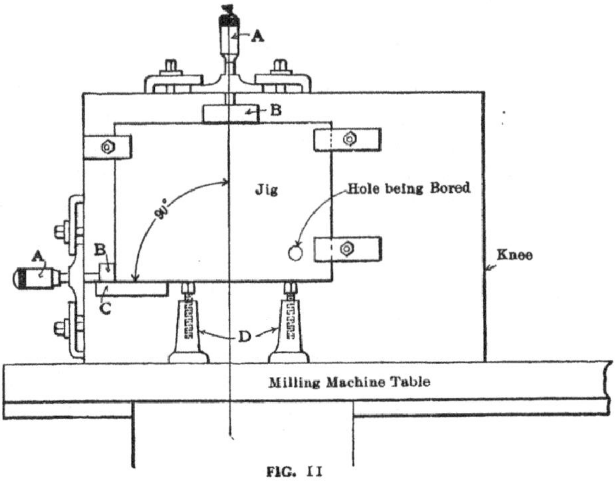

FIG. 11

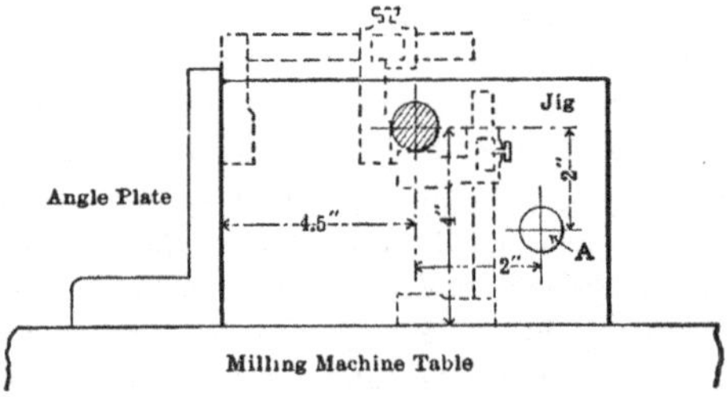

FIG. 12

OTHER WAYS OF LOCATING JIGS ON THE MILLER

of standard size blocks 1-, 2-, 3-inch, etc., a variety of jig work can be conveniently handled. The table and knee of the miller are clamped fast and the work adjusted properly on the face of the angle plate by the micrometer gages and size blocks. The scheme will be clear from Fig. 11, where the depth gages are shown at *A A* and the size blocks at *B B*, with a parallel *C* placed under the lower edge of the jig, to support the lower size block. The work is supported as indicated, by jacks *D*.

THE VERNIER ON THE MILLING MACHINE

The vernier is frequently used on the miller for the production of accurate jig work. The method consists briefly in clamping the scale or beam of the vernier to the table, and the vernier plate to a bracket fastened to the saddle, the table being accurately set for the different holes to be bored by direct reading of the vernier. For vertical adjustment of the work, another scale may be attached to the face of the column and the vernier plate to the knee. The scale for the table is best attached by adjustable studs fitted to the T-slot of the front edge which carries the trip dogs for the feed; the vernier plate should be mounted on a block which will admit of adjustment on the bracket attached to the knee.

In this connection it should be stated that the Brown & Sharpe Manufacturing Company makes for this purpose a special scale and vernier; the scale, which is 24 inches long, may be conveniently attached to the front of the table, and the vernier plate accompanying this scale has a holder which is readily secured to the saddle when the device is required for jig work. Further illustration of the use of the vernier on the miller is presented in Chapter III.

THE VERNIER HIGHT GAGE ON THE MILLER

The vernier hight gage may be used for setting the work for boring in the miller, in combination with a plug or arbor in the spindle. A convenient size of plug is 1 inch, and it must necessarily be true and straight. The jig is secured to the table with one end against an angle plate, as shown in Fig. 12. Now if one hole is to be 4.500 inches from the end of the jig and 4 inches from the bottom, the hight gage is set to 4 inches — ½ inch (i.e., one-half diameter of plug), or 3½ inches, and the table elevated until the gage will just slide under the 1-inch plug, as shown in the sketch by dotted lines. Then the hight gage is set to 4½ inches — ½ inch, or 4 inches, and the table moved until the gage when rested against the angle plate will just contact with the side of the plug. The setting is then correct for the first hole; after this is bored to size, the hight gage is set to 4 inches — 2 inches — ½ inch, or 1½ inches, and the knee elevated until the gage when placed on the table will again just contact with the under side of the plug, thus giving the correct vertical position for boring the second hole *A*. The longitudinal adjustment is obtained by setting the hight gage to 4½ inches + 2 inches − ½ inch, or 6 inches, and moving the table until the gage will once more contact with the face of the angle plate and the side of the 1-inch plug. The work is now in the correct position for the boring of the hole *A*.

By this method, no matter how many holes are to be bored, all measurements may be taken from the surface of the table and the face of the angle plate, thus obviating the possibility of added errors in the work.

THE DISK METHOD

For locating holes in small jigs the disk method is a favorable one, especially with watch-tool toolmakers, and is considered one of the most suitable of methods where a number of holes have to be bored with their centers within a limit of 0.0002 inch.

The center distances being known, disks and plugs of such diameters are made that with their outer diameters in contact with two other disks or plugs their centers will be a continuation of the axes of the holes to be bored. The disks should be first roughed out and then by means of solder fastened to a solder chuck on a bench lathe, as in Fig. 13 and the outer face carefully squared off, the diameter ground to exact size and a sharp V center put into the front face.

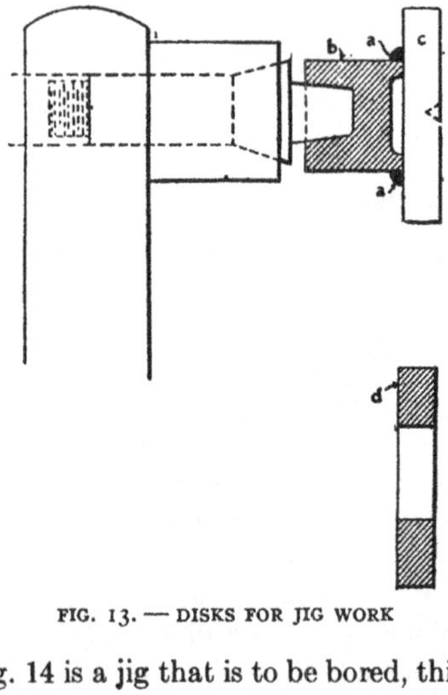

FIG. 13. — DISKS FOR JIG WORK

Consider that Fig. 14 is a jig that is to be bored, this jig having a center hole, a hole 1 inch from the center and another hole 1.1 inches from the center, the radial center lines of the two outer holes forming an angle of 45 degrees. It will be found convenient to bore the center hole first and then form a recess in the face of the jig, as in Fig. 15. Say this recess is bored out exactly 2.85 inches diameter; for locating the hole B a disk B' is made precisely 0.85 inch diameter, placed in contact with the rim of the jig and secured to the jig face with a few drops of shellac or wax around its edge. The jig may then be indicated upon the lathe face plate by means of a sensitive indicator resting in the V center in the disk, the disk removed and the hole B bored. The size of this disk is of course obtained by subtracting center distance *b*, or 1 inch, from the recess radius, 1.425, giving distance *c*, or 0.425, which multiplied by 2 is the diameter of the disk.

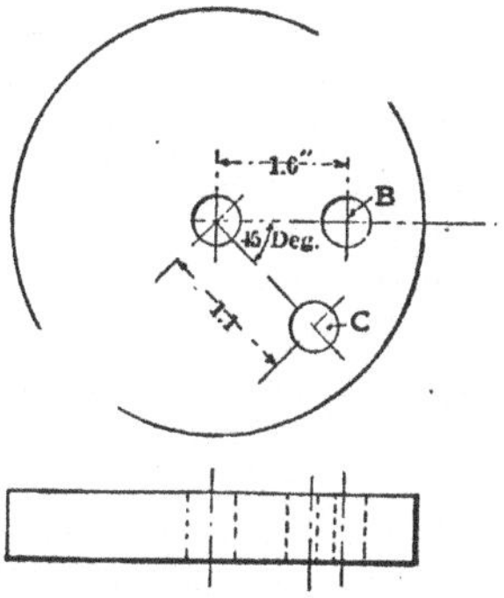

FIG. 14

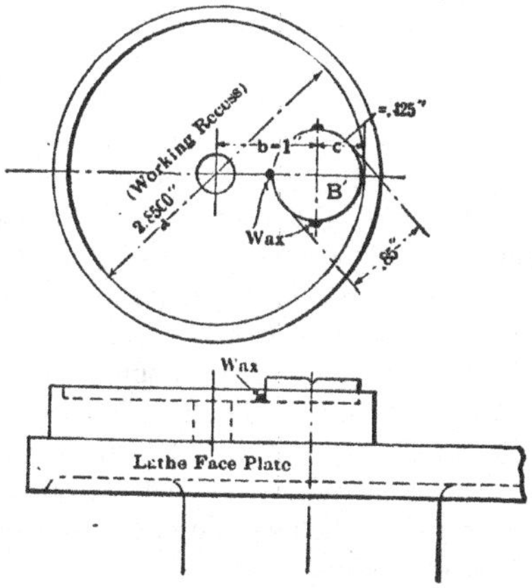

FIG. 15

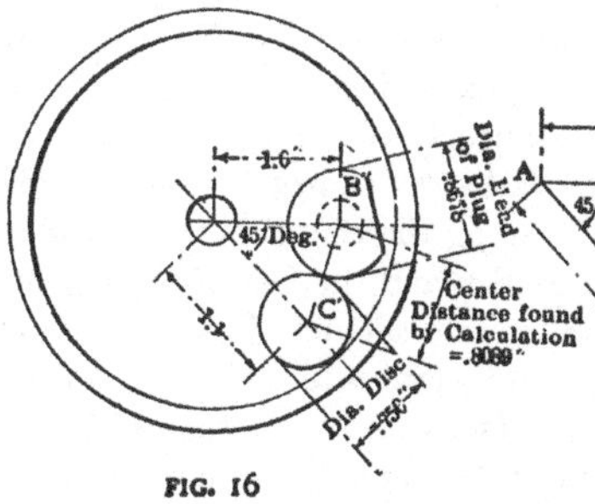

FIG. 16

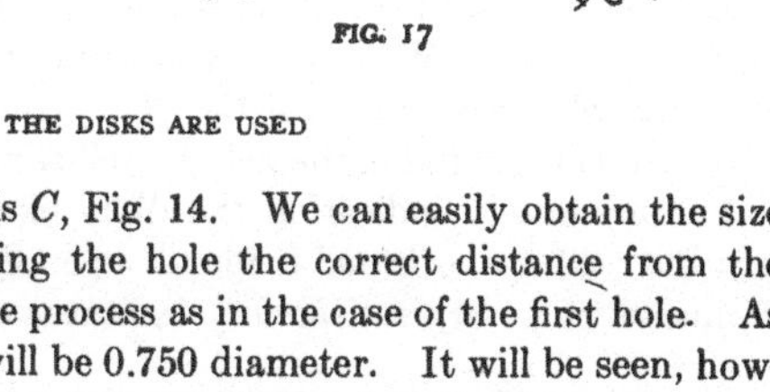

FIG. 17

HOW THE DISKS ARE USED

The next hole to be bored is *C*, Fig. 14. We can easily obtain the size of the disk required for locating the hole the correct distance from the center by using the same simple process as in the case of the first hole. As indicated in Fig. 16, the disk will be 0.750 diameter. It will be seen, however, that to locate the disk correctly in relation to hole *B* we must first find the center distance between *B* and *C*, Fig. 14. This is readily done by simple trigonometrical calculation as follows: Referring to the diagram, Fig. 17, we have an oblique-angle triangle with sides representing the center lines drawn through the three holes in the jig. Sides *A B* and *A C* are 1 inch and 1.1 inches respectively, and we have to find side *B C* from these two sides and from angle *A*, which is 45 degrees. If we draw a perpendicular, as *C D*, we divide the triangle into two right-angle triangles and can then find $A\,D$ by multiplying 1.1 by cosine 45 degrees, or $1.1 \times 0.7071 = 0.7778$; 1 inch $- 0.7778 = 0.2222$ inch $=$ length of line $D\,B$. Next we find the length of $C\,D =$ sine 45 degrees or $0.7071 \times 1.1 = 0.7778$. We now have two sides of triangle $C\,D\,B$, and dividing $D\,B$ (0.2222) by $C\,D$ (0.7778) we have $0.2856 =$ tangent 15 degrees 56 minutes, which is the angle formed by lines $C\,D$ and $C\,B$. The secant of this angle is 1.0399, which multiplied by 0.7778 inch $= 0.8089$ inch, or the length of line $C\,B$. This is then the correct distance between the centers of holes B and C of the jig.

The disk C', Fig. 16, for hole C has been made 0.750 inch diameter in order to get the right distance from the jig center for that hole; half this diameter, or 0.3750 inch, is subtracted from 0.8089, the figured distance between centers B and C, giving 0.4339 inch as the radius of the head of a plug B'' for hole B. This plug has head and body finished exactly concentric, having the body fitting nicely in hole B and the head, whose diameter is 2×0.4339, or 0.8678 inch, being cut away at one side to clear the rim on the jig. Disk C', when brought into contact with the plug head and the jig rim, is secured with a few drops of shellac, and the jig is again trued on the face plate by the indicator, the disk removed and hole C drilled and bored.

For holding the disk to the jig some prefer a mixture of beeswax and shellac in equal parts, and others use beeswax and rosin. After melting the shellac or the mixture in a pot, it is applied to the disk by a wire.

In mounting the disks on the bench lathe, as shown in Fig. 13, for finishing to the required diameters and forming the center hole, no solder must be permitted between the face of the disk and the face of the chuck. Instead, the solder is dropped at several points around the edge, as indicated at *a a*. The chuck has a brass front plate *b* to which the solder adheres quite firmly; this plate is of course faced off true before soldering the disk in place. While the usual form of disk is that represented at *c*, for extremely accurate work the disk is made in the form shown at *d*, hardened, and ground in the hole as well as on the periphery.

For accurately indicating the disks like *c*, a sensitive device with a light but stiff arm of wood is found superior to any other indicator. Fig. 18 illustrates a very satisfactory indicator and is self-explanatory.

POSSIBILITIES OF THE DISK METHOD

It will appear that the disk method can be applied to an endless variety of nice jig work where center distances are short and disks of unduly large diameter therefore not required. It is obvious that the recess turned in the face of the jig, Figs. 14 to 16, can be of any convenient diameter so long as it allows sufficient room for the disks for locating the holes farthest from the center.

In many cases the disks may be located conveniently without the aid of the outer rim formed on the jig in Fig. 15. If we have given the center distances for three holes to be bored as in Fig. 19, we can make three disks of the proper size, place their edges in contact, secure them to the jig with wax or shellac and indicate and bore one hole after another. In the case shown in Fig. 19, where the distance from center to center is equal, the disks are of course all made to one diameter, equal exactly to the center distance between the holes, as shown in Fig. 20.

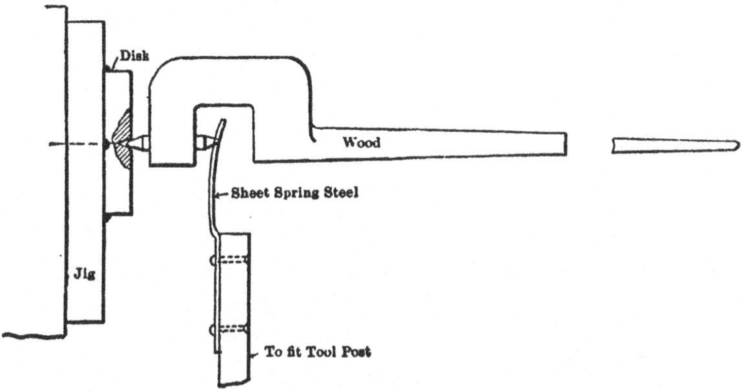

FIG. 18. — INDICATOR USED WITH THE DISKS

Where the center distances desired are all different, as in Fig. 21, the diameters of the disks may be easily found, as suggested by the diagram in Fig. 22.

Let x = half diameter of disk A
y = half diameter of disk B
z = half diameter of disk C

According to the dimensions in Figs. 21 and 22,

$$x + y = 0.765$$
$$x + z = 0.710$$
$$y + z = 0.850$$

If $x + y = 0.765$
and $x + z = 0.710$

Subtracting, we have $y - z = 0.055$

Now $y + z = 0.850$
and $y - z = 0.055$

Adding, we have $2y = 0.905$, diameter of disk B.

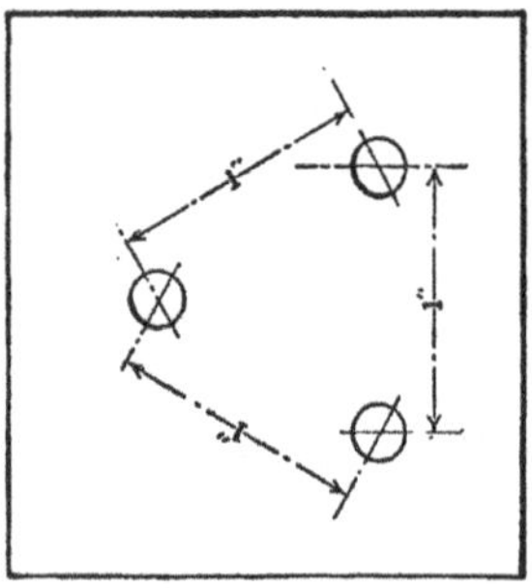

FIG. 19

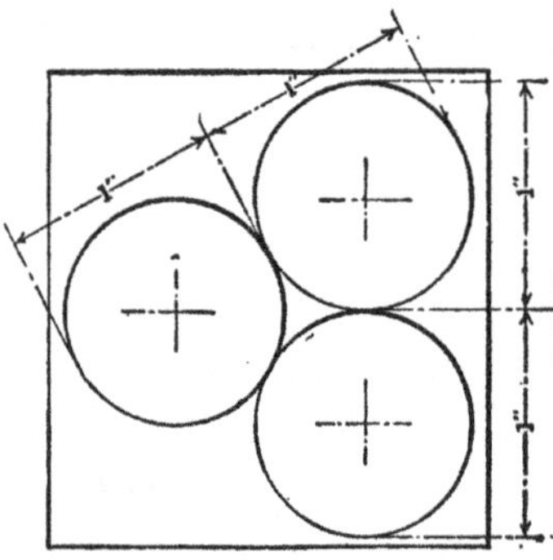

FIG. 20

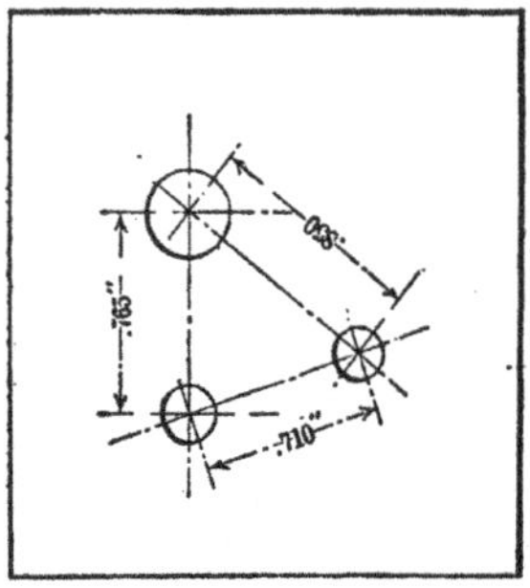

FIG. 21

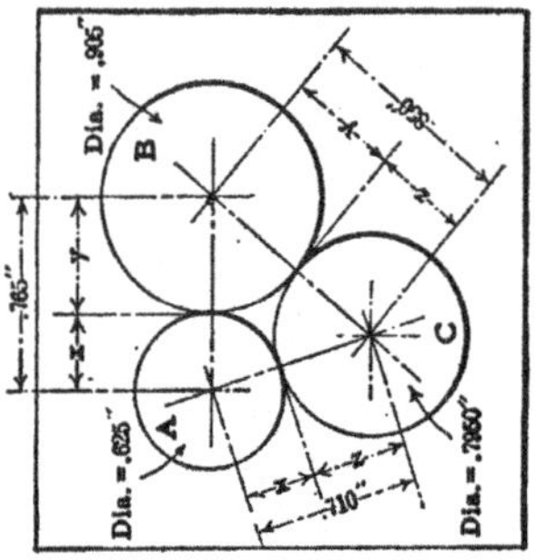

FIG. 22

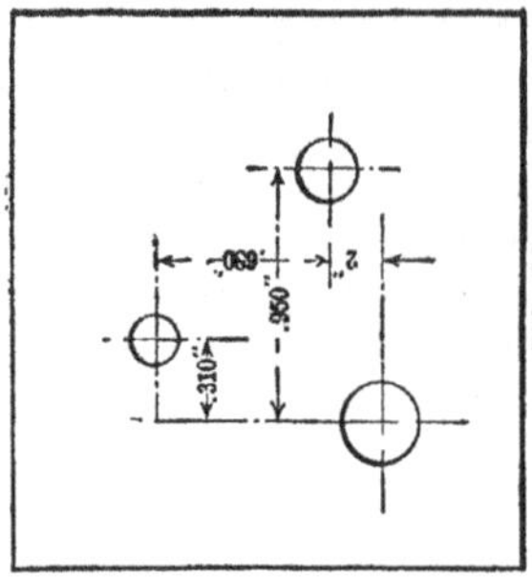

FIG. 23

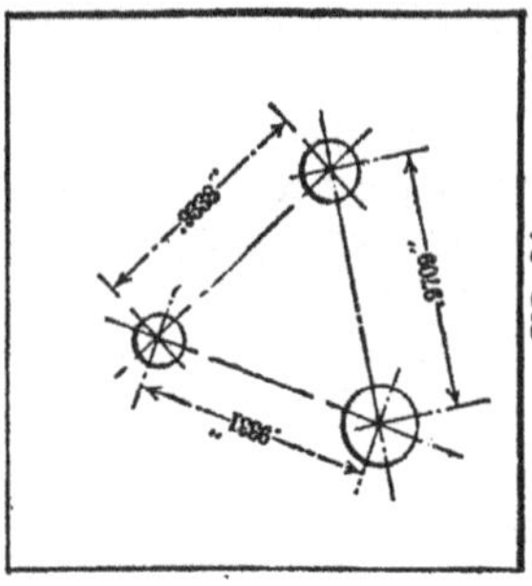

FIG. 24

APPLICATIONS OF THE DISK SCHEME

Therefore $y = 0.4525$

As distance $x + y = 0.765$

$x = 0.765 - 0.4525 = 0.3125.$

Multiplying by $2 = 0.625$, diameter of disk A.

To get the diameter of disk C, subtract from 0.710, which is equivalent to $x + z$, the value of x, or 0.3125, leaving 0.3975, or the value of z, which multiplied by 2 gives 0.7950 as the diameter of the third disk.

When the dimensions between centers are given, as in Fig. 23, from horizontal and vertical center lines, a simple calculation in square root enables the direct distances from center to center to be obtained, each of these center distances representing the hypotenuse of a right-angle triangle whose base and hight are given on the drawing. After finding the direct center distances which are placed on the sketch in Fig. 24, the disk sizes may be determined in the way just described in connection with Figs. 21 and 22.

These few illustrations will give some idea of the possibilities with the disk scheme on tool work. The applications are too numerous to be more than merely suggested here, especially when combinations of four disks, or combinations of disks and size blocks, are considered.

MASTER-PLATE OR TRANSFER METHOD

When either a model or a jig for one member which must match up with another has been made, the scheme of doweling the jig that is to be bored to the model or other jig is of great value in locating the holes. While this idea is known to some as the transfer scheme, it is practically the same as what is often called the master-plate method, which is fully described in Chapters IV, V, and VI.

Briefly, the system is to make plugs fitting the spindle of the lathe with tits fitting the holes in the model, the tit being finished to size after placing the plug on the lathe, and the model being doweled to the jig that is to be bored, as in Fig. 25. There is obviously a certainty of perfect duplication of center distances with this scheme. The idea is applicable to both the lathe and the milling machine.

In designing jigs from models, from other jigs or from master plates, it is advisable to make the jig in flat sections or plates, which permit of much easier handling than heavy box castings.

SPECIAL METHODS — ADJUSTABLE JIGS

Adjustable jigs for drilling other jigs are not used very commonly, although there are some cases where such tools are exceedingly handy. Fig. 26 illustrates a jig of this class by which holes may be located readily in simple plate jigs, this tool being conveniently used in connection with the vertical-spindle attachment on the milling machine. The ledges *A A* on the flat true plate *B* are finished exactly 1 inch thick. The bushing *C*, which is adapted to receive drill bushings of various sizes, is exactly 2 inches diameter. The adjustment to the desired position for putting a hole through the jig which is to be made is obtained nicely by measuring with micrometers over the outside of the bushing *C* and the outer edges of *A A*, as indicated; or when more convenient to do so, between the inside of the bushing and the inner faces of *A A*. Allowance is readily made

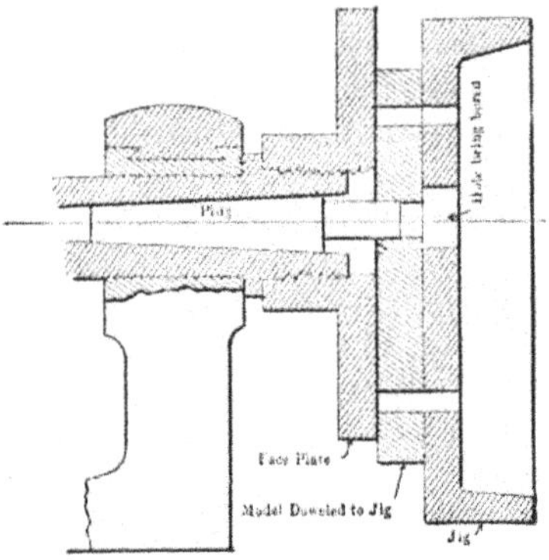

FIG. 25. — USING THE MASTER PLATE

when measuring for half the diameter of the bushing and the thickness of *A*, as each of these dimensions is just 1 inch.

SPECIAL METHODS WITH LARGE JIGS

In making large jigs for side plates of printing press frames, etc., the work is frequently done under a drill press. Fig. 27 will give the reader an idea of how such work may be handled.

D is a movable hardened and ground drill guide which may be clamped on the jig after having been properly located from the edges of knees *E* and *F*.

The enlarged view of the drill guide *D* will possibly better illustrate this device. The base, outer diameters and hole are carefully ground and are square with each other. End-measure plugs or vernier calipers are used in locating the guide; when this is located and clamped, the hole in the jig is carefully spotted and drilled and reamed with tools passing through the drill guide. Where all the holes to be bored are not of one size, bushings to suit the tools are made to fit the guide. When the drill guide cannot be otherwise clamped, it is customary to drill and tap two binding-screw holes directly into the jig plate.

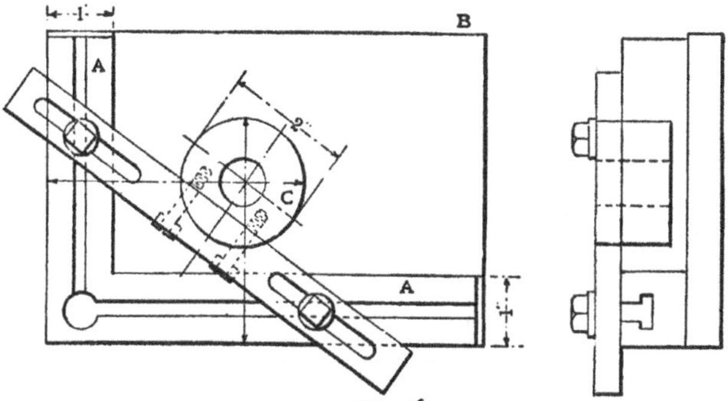

FIG. 26

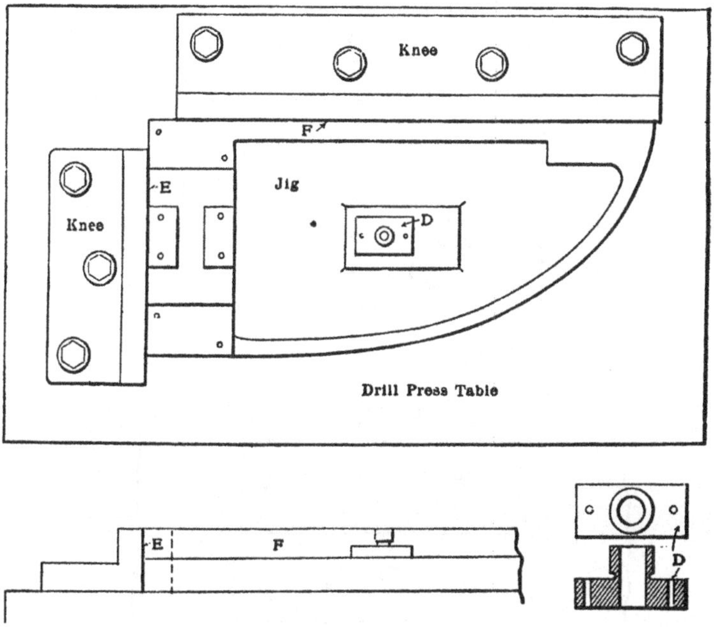

FIG. 27

SPECIAL JIG METHODS

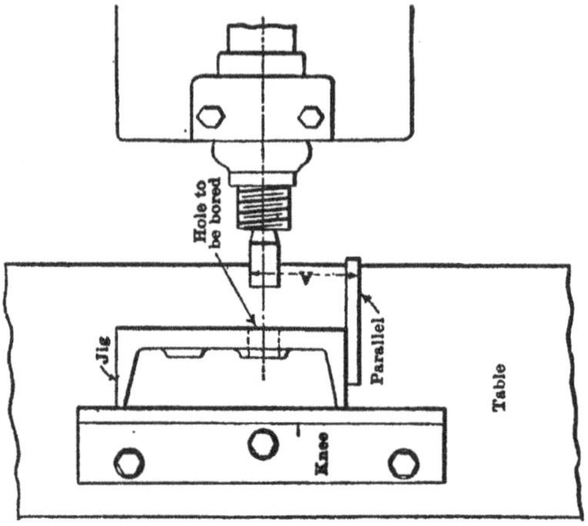

FIG. 28

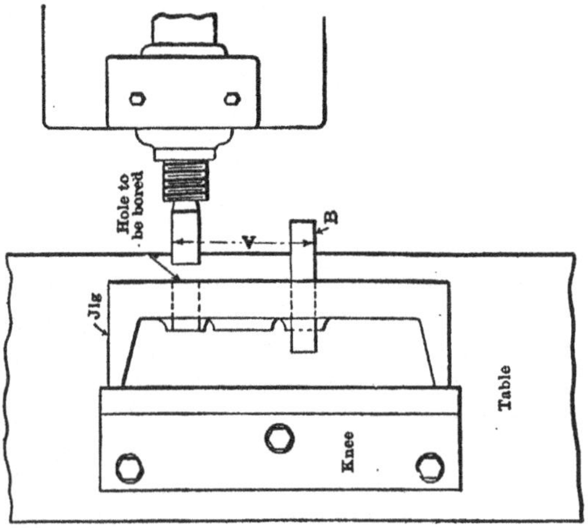

FIG. 29

SETTING JIGS BY PLUGS AND ARBORS

LARGE WORK ON THE HORIZONTAL BORING MACHINE

Only the drill press, lathe and milling machine thus far have been referred to in connection with the boring of drill jigs. The horizontal boring mill is also used for box jigs too large to be handled on the milling machine. It is customary on the boring mill as well as on the milling machine to take measurements from two edges of the jig to a true arbor held in the head spindle, as indicated by Fig. 28. A parallel can be secured at the end of the jig and a micrometer used to measure, as at *A*, over the parallel and arbor, allowance being made for the thickness of the parallel and half the diameter of the arbor. The distance from the upper edge of the jig may be obtained in similar fashion.

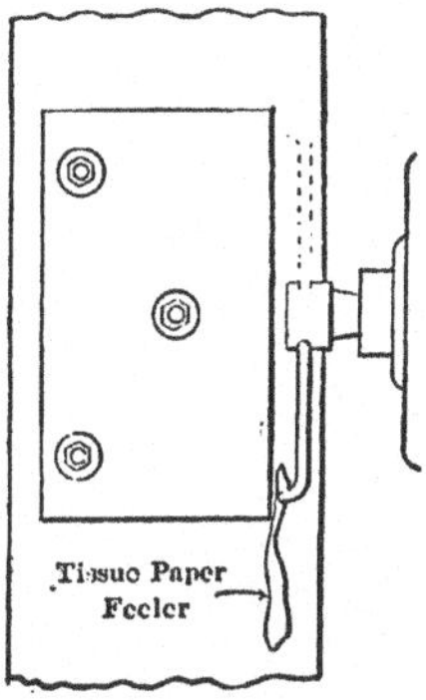

FIG. 30. — SETTING A KNEE

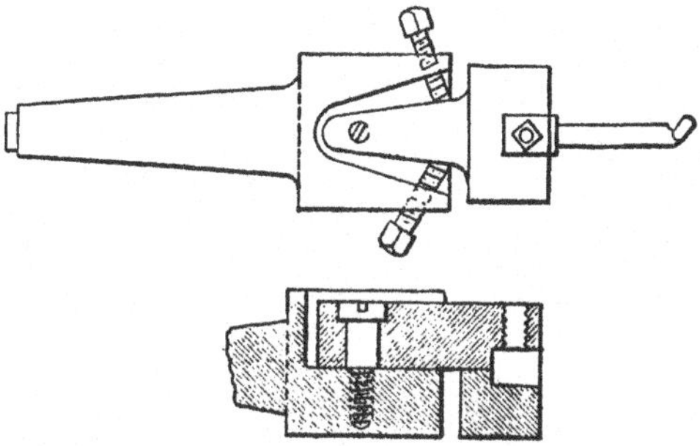

FIG. 31. — A BORING TOOL

In some cases, after one hole has been bored, the jig is located for subsequent holes from a plug *B*, Fig. 29, snugly fitted in the hole already bored. Here again measurements may be taken by a micrometer over the plugs in the jig and the one in the spindle.

For setting an angle plate or knee square with the spindle on a horizontal boring mill or miller, the swinging tramming bar and arbor shown in Fig. 30 will be found useful. The tool is first swung to one side and then to the other of the knee, a tissue-paper feeler being used between the tram point and the knee. With this method a perfect setting of the knee is quickly secured.

It may be of interest at this point to call attention to an adjustable boring tool which is useful in either the milling machine or horizontal mill for boring holes in jigs and fixtures. The tool is shown in Fig. 31 and will be understood without any description.

WORK ON THE DIVIDING HEAD

Reference should be made here to the use of the dividing head on the miller in handling certain classes of jigs. One application of this device is found in the indexing of circular jigs which require a series of holes spaced about one or more circles or along an arc of a circle. The accuracy of the work may be tested by figuring the chord between adjacent holes and then measuring with micrometers over plugs fitted in the holes. If inaccuracies in center distances are found after two holes are bored, corrections can be made by rotating the work slightly and reboring until the measurement over the pair of plugs is correct.

CHAPTER II

Locating and Boring Oblique Holes in Jigs

A type of jig which the toolmaker frequently has to build, is represented in Fig. 32.

The jig body and plate are first carefully planed on all surfaces as indicated by the finish marks, using ordinary methods and taking particular

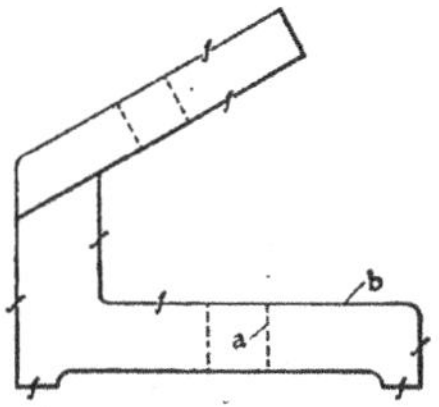

FIG. 32. — THE JIG

care to insure the base and sides being square and at right angles with each other. The accurate planing of the angular face of the jig body, to which the plate is attached, may be considered the next operation, and this may be accomplished by several different methods.

SETTING WORK FOR PLANING

Fig. 33 illustrates an excellent way to position a plate correctly so as to plane or mill any desired angle accurately. Briefly, the method is to

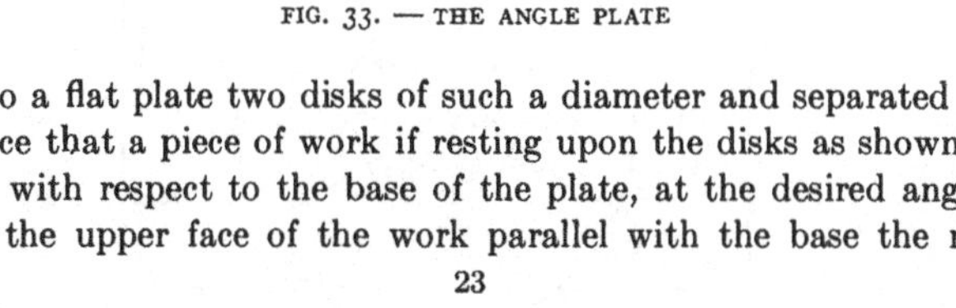

FIG. 33. — THE ANGLE PLATE

attach to a flat plate two disks of such a diameter and separated at such a distance that a piece of work if resting upon the disks as shown will be inclined with respect to the base of the plate, at the desired angle. By planing the upper face of the work parallel with the base the required

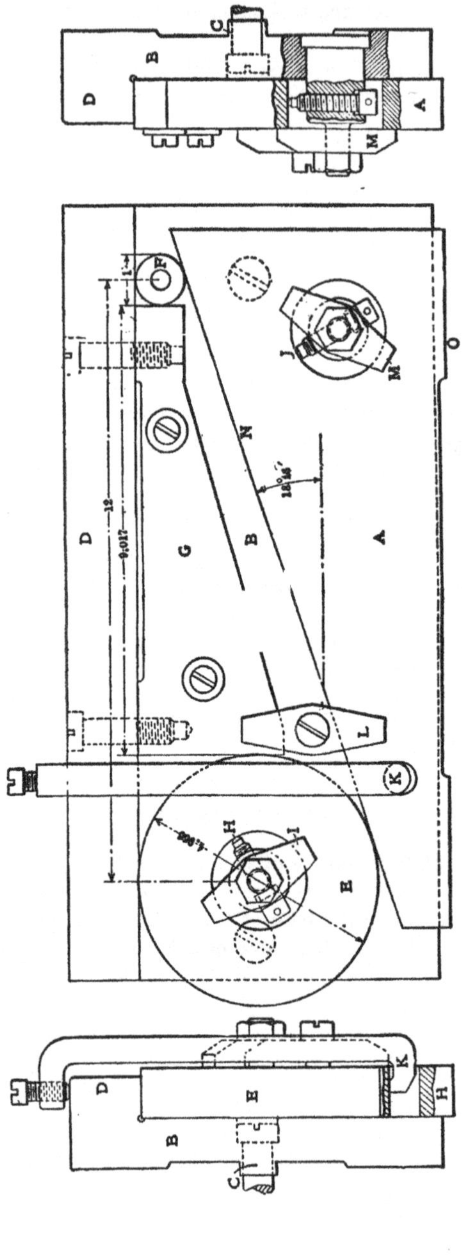

FIG. 34. — PLANING DEVICE FOR MASTER ANGLE

angle will be produced, assuming the lower edges of the two disks are parallel with the base of the plate.

MAKING MASTER ANGLE PLATE[1]

It may be of interest to consider here in detail the method of making such a plate. Let *A*, Fig. 34, be the required angle plate. *B* is a plate tongued at *C* to fit the groove in the planer table, and having at one side a raised rib *D*. *E* and *F* are disks and *G* a separator to locate the disks the required distance apart. The separator *G* is attached to the rib *D* by two screws. The large disk *E* is held in place against the rib *D* and the end of separator *G* by the screw *H*, and held down by clamp *I*. The small disk *F* is held against rib and separator by the slanting edge of the angle plate. The angle plate *A* is drawn against the disks by the screw *J* and clamp *K*, and held down by clamps L and *M*.

In making the angle plate shown at *A*, it was planed all over, the edge *N* scraped, then carefully placed and drawn against the disks *E* and *F* and clamped as described. The finishing cut was then made on edge *O*, which was recessed at the center. Just before taking this final cut, a light cut was made on both edges of the rib *D;* these edges being thus exactly parallel furnished a means of verification of the angle plate at any future time, by replacing it in position and measuring over all with a micrometer.

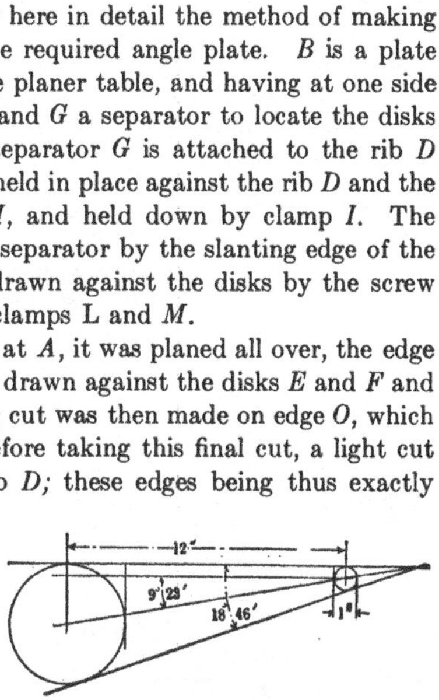

FIG. 35. — THE GEOMETRY OF THE PROBLEM

The diameter of the disks and length of separator were calculated as follows: 18 degrees 46 minutes was the required angle; 12 inches was assumed as a convenient distance between centers of disks, and 1 inch diameter for small disk. Referring to Fig. 35,

$$\frac{18^\circ\ 46'}{2} = 9^\circ\ 23'$$

$$\begin{array}{ll}
(\tan.\ 9^\circ\ 23' = 0.16525) \times 12 = & 1.983 \\
+ \text{ radius of small disk,} & 0.50 \\
\hline
= \text{ radius of large disk,} & 2.483 \\
\quad \text{diameter of large disk,} & 4.966 \\
\quad \text{Distance between centers} = & 12. \\
- \text{ added radii,} & 2.983 \\
\hline
= \text{ length of separator,} & 9.017
\end{array}$$

[1] "Making Master Angle Plate," contributed to the *American Machinist* by Jas. Dangerfield.

THE ACCURACY OF THE PROCESS

The accuracy of the foregoing procedure is only limited by the expertness of the toolmaker, assuming the disk diameters and center distance are correctly calculated. It is obvious that the angular face of the jig body may be planed directly by holding the jig in the manner described, but in most cases it will be more convenient to produce a master angle plate by this method and to utilize the latter in correctly locating the jig body when planing the inclined face and also when boring the hole for the drill bushing.

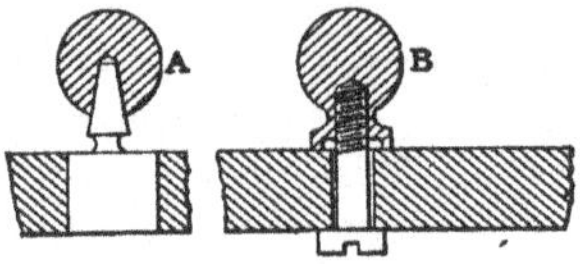

FIG. 37

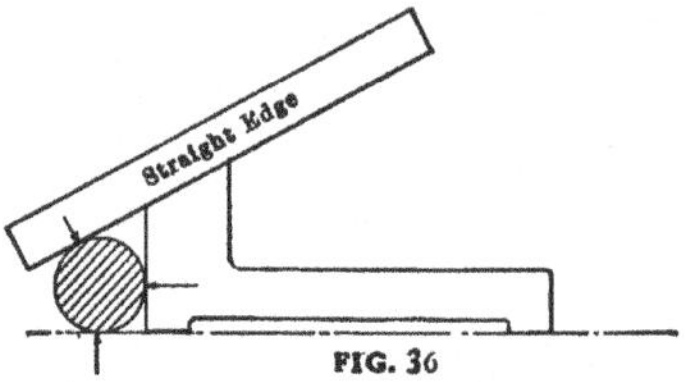

FIG. 36

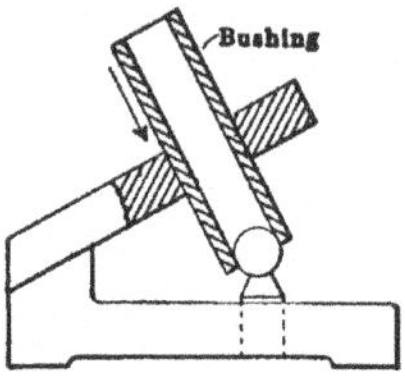

FIG. 39

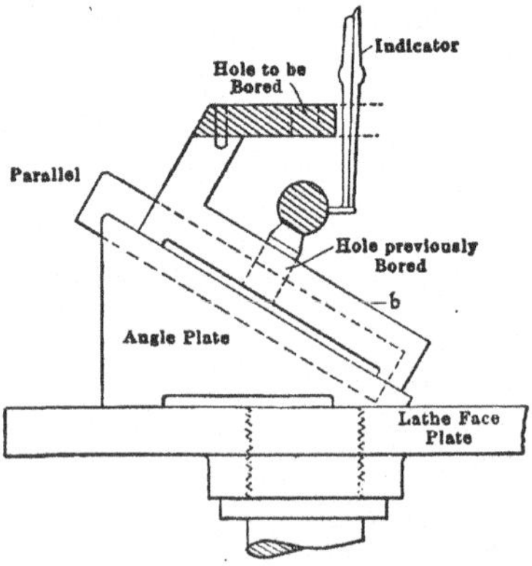

FIG. 38

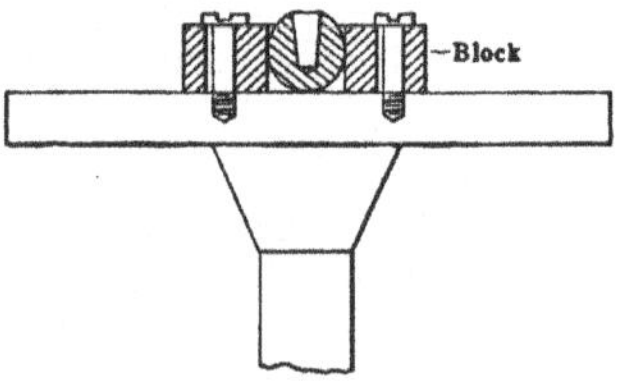

FIG. 40

LOCATING AND BORING OBLIQUE HOLES

The application of cylindrical disks when planing the inclined face of a jig to a correct vertical measurement from the base or other surface and horizontally from the end of the jig body is illustrated by Fig. 36.

LOCATING THE WORK FOR BORING

Assuming that the jig body and plate have been planed as outlined above; the next step may be to bore hole *a* shown in Fig. 32, various methods for which have been described in these columns. The oblique hole in the plate may be bored after the plate is attached to the body or previously. In the event of its being considered advisable to perform the boring operation with the plate assembled, it is first necessary so to position the jig on the lathe face plate (assuming a lathe is to be used for the boring operation) as to insure the inclined hole being bored at the proper angle and distance from the first hole bored in the jig body. The master angle plate which has been prepared will give the required angular setting, and a spherical button in connection with a test indicator as shown by Figs. 37 and 38 will afford a means of properly locating the jig for the correct position of the hole. One edge of the jig base should rest against a parallel which is clamped to the edge of the master angle, and the whole is securely clamped to the face plate when the indicator point is stationary. Assuming that the intersection of the axis of the two holes should be 0.625 inch from the face *b*, and also assuming that the spherical button is 0.5 inch in diameter, it is obvious that the end of the button should be 0.625 inch $\times \frac{.5000}{2}$ from face *b* in order that the center of the button may be at the intersection of the axes of the two holes. The button if so positioned and being a perfect sphere affording a continuous circular path for the indicator at all cross sections will thus be found an absolutely correct means in connection with the indicator of properly positioning the work.

LOCATING THE JIG PLATE AFTER BORING

Should it be considered advisable to bore the hole in the plate before assembling the two members, the ball button may be used in connection with a cylindrical locating bushing as shown in Fig. 39 for properly positioning the two members before doweling together.

Ball buttons of various diameters are used, but on the ordinary run of work a diameter of 0.5 inch will be found convenient. Various methods of making accurate balls have been described, but it will be found entirely satisfactory to use commercial balls that are practically perfect spheres. A central conical hole should be bored in one side and the ball then mounted on a stem as shown at *A*, Fig. 37. One method of boring the hole central is to clamp a block on the face plate of a good lathe and carefully bore a hole in the block equal to the diameter of the ball, then push in the ball and either shellac or clamp the ball in place while boring the hole. Fig. 40 illustrates ball and holder.

RESTING THE WORK ON PLUGS AND HIGHT BLOCKS

A variation from the preceding method which permits of dispensing with the angle plate is to bore four holes, for plugs, in the frame of the jig, as shown by Fig. 41. The plugs, which are temporarily used in place of the angle plate, must be of such diameter as to incline the jig to the proper angle. When planing and boring, the jig is supported by the plugs for the angular work. If the plugs are all of one diameter they should rest on hight blocks of suitable dimensions to produce results desired, as indicated by Fig. 42. In either case it is of course necessary to position the four holes carefully and to calculate the dimensions of plugs and hight blocks accurately, and if no error has been made correct results should be obtained.

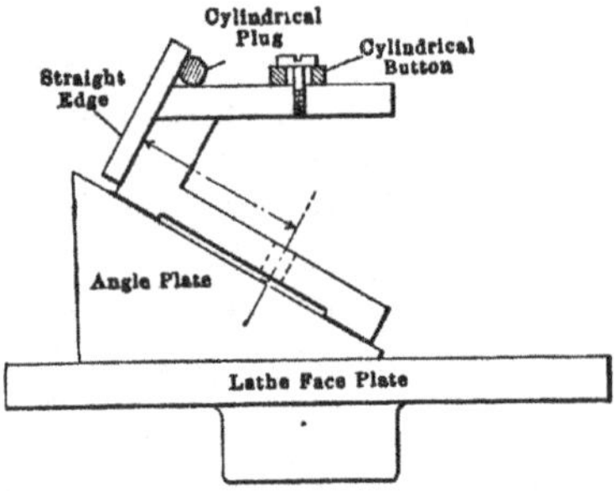

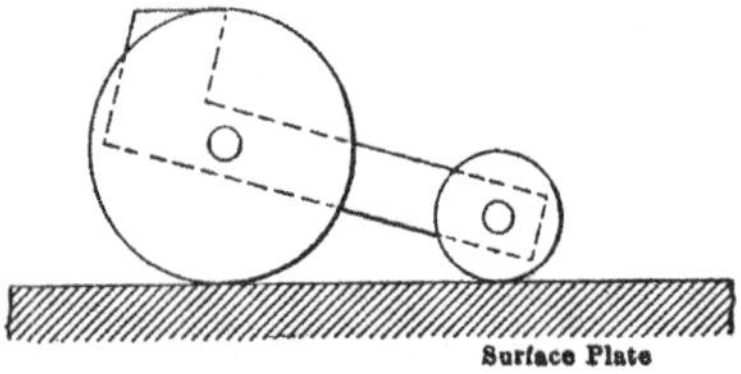

FIG. 41

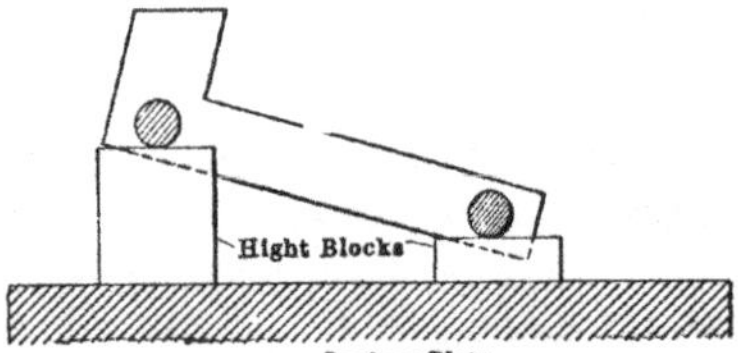

FIG. 42

SETTING ANGULAR WORK BY MEANS OF PLUGS

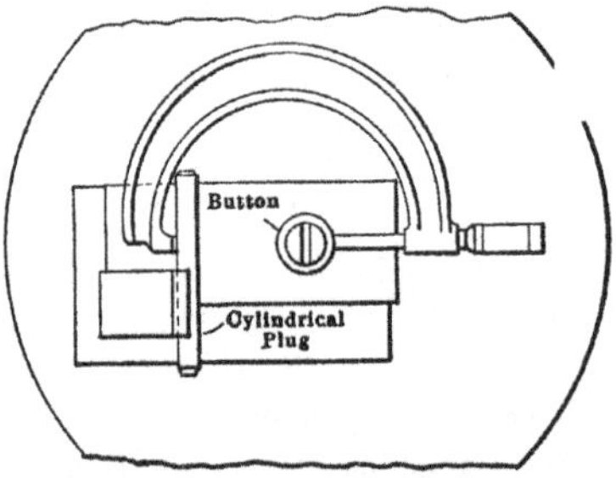

FIG. 43. — USE OF MICROMETERS AND BUTTONS

Still another method of locating a jig of the type described, previously to boring the oblique hole, and which does not require the use of a ball button, is first to locate properly by micrometer measurement an ordinary indicating button or disk as shown by Fig. 43 in connection with a parallel and cylindrical plug, and then indicate the work and clamp by ordinary methods. The master angle may be used to obtain the correct angular relation.

USE OF THE MILLER TABLE

The universal milling machine with its graduated swivel saddle affords an excellent means of boring angular holes and of milling angular faces. In very accurate work the milling machine table may be properly inclined by means of master angles or by parallels having disks of varying diameters positioned so as to produce desired results. In some cases the use of protractors is found the most suitable method of locating the table and work. Fig. 44 illustrates a piece of work having two holes at an angle; the successive steps in boring by means of the universal milling machine and its graduated swivel saddle being shown by Figs. 45, 46 and 47. To the top of the table is clamped a disk so positioned that its axis is a continuation of the axis of rotation of the saddle.

The accurate positioning of this disk may be accomplished by clamping a sensitive indicator to some stationary point and adjusting the disk, and partially rotating the table together with the disk until continuous contact between indicator and periphery of the disk shows no movement of the long arm of the indicator. The disk being of known diameter affords means by which the correct longitudinal movement of the table may be calculated. For obtaining the desired longitudinal adjustment the use of a vernier attachment or depth micrometer will be found satisfactory for accurate work. Some prefer size blocks. On work of ordinary accuracy the regular screw and micrometer dial are commonly used although the back lash must always be taken up in the same direction. The steps necessary to take, and the entire scheme, are self-explanatory from the figures referred to. It may be mentioned, however, that in many cases the disk may be made of a diameter that will permit the work to be so positioned with respect to the axis of the saddle that only swinging movement to the table will be required, thus eliminating the longitudinal movement altogether.

OTHER METHODS

Still another method that is often found satisfactory when locating and boring holes in jigs of the type indicated by Fig. 44 is to use two spherical buttons attached to the jig as shown by Fig. 48, and to indicate the work from the buttons in a lathe as in Fig. 49.

The indexing or dividing head of the universal milling machine is very commonly used when jigs like Fig. 50 are to be bored. The work in this case may be mounted on an arbor between the dividing head centers and indexed around to give the correct angular position of the various holes.

FIG. 44

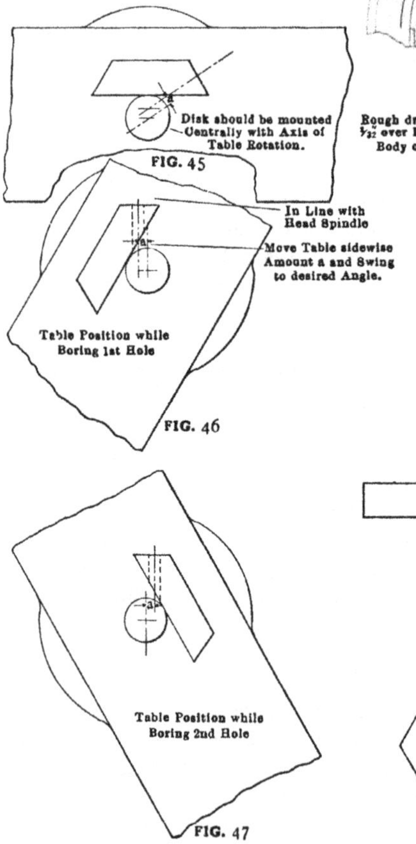

FIG. 45

FIG. 46

FIG. 47

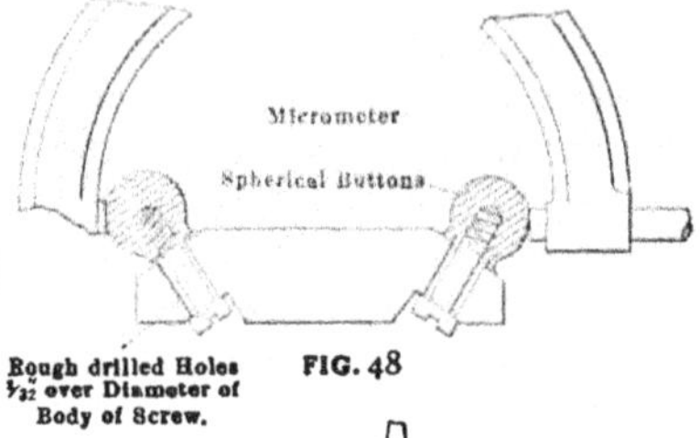

FIG. 48

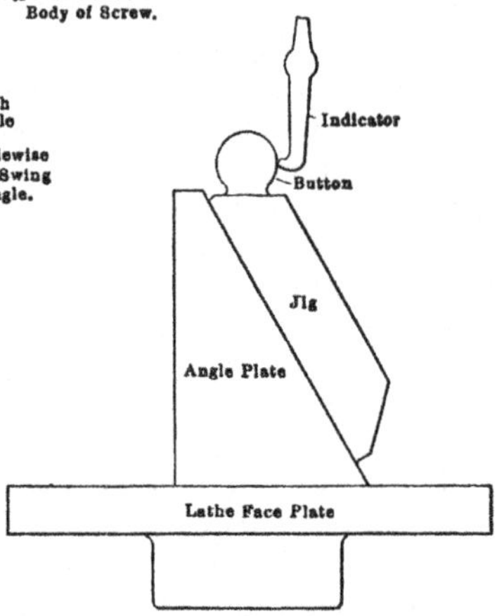

FIG. 49

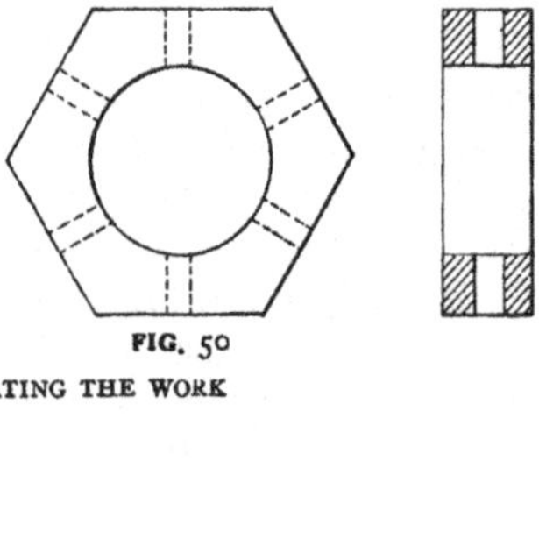

FIG. 50

OTHER METHODS OF LOCATING THE WORK

CHAPTER III

ECONOMICAL JIG WORK ON THE MILLING MACHINE[1]

THE value of the milling machine in the production of jigs is appreciated by those who have large quantities to handle and whose limit of accuracy will allow of 0.001 inch variation, which under average conditions and with average men is about the closest practical limit maintained. Some shops do on the miller a great deal of the work on the jigs ordinarily done on the planer or shaper, such as surfacing the feet, bosses and locating points and cutting slots. Others use it principally for drilling and boring pin and bushing holes which have to be accurate. On both classes of work the milling machine shows excellent results on small and medium-sized jigs. For this purpose a machine must be kept in first-class shape and better care taken of it than toolmakers ordinarily take of their machines. On that account it is preferable that the machine and tools be handled by one man and that he be held responsible fcr their condition.

THE TOOL EQUIPMENT

The equipment consists of a vertical attachment and an adapter, Fig. 51, for the table for holding the work; an assorted collection of end, side and facing mills for regular milling operations; verniers attached to the table and knee; and a full outfit of spotting drills, boring bars and reamers for putting in holes. An overhanging arm having a bushing for supporting the arbor is preferable to one with a center, as the bushing when supporting a boring bar can be taken out and the bar removed without disturbing the work of the adjustment of the arm.

The adapter for the table should be made with a tongue on the bottom fitting the slot in the table accurately, and T-slots on the top running lengthways and crossways at frequent intervals, as shown, those running lengthways being parallel to and those crossways square with the tongue on the bottom. The slots should be made to take standard bolts, and the narrow part should be the same width in all, so that standard blocks can be inserted to facilitate setting the work. Having this plate fastened on the table carefully, as in Fig. 51, the slots should be squared up from the spindle by tramming carefully.

[1] Contributed to the *American Machinist* by E. A. Johnson.

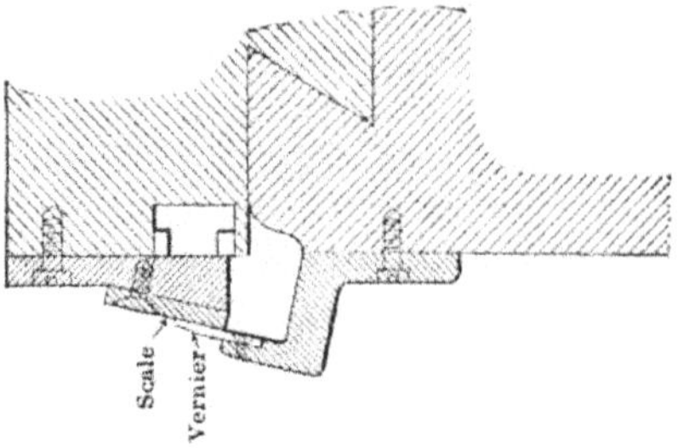

FIG. 52

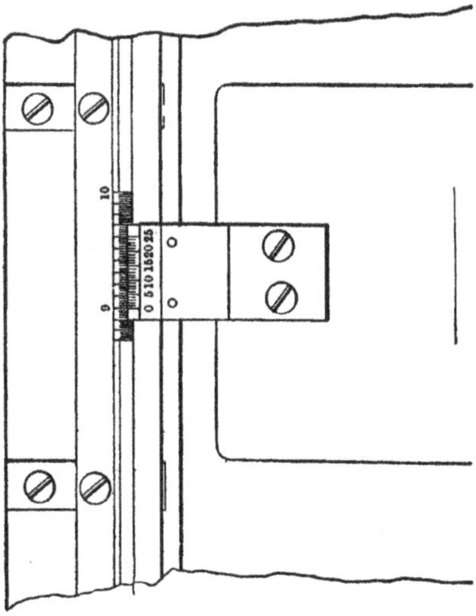

EQUIPPING THE MILLER FOR JIG BORING

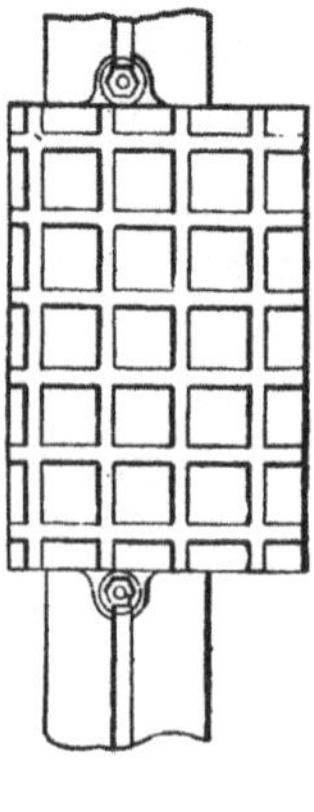

FIG. 51

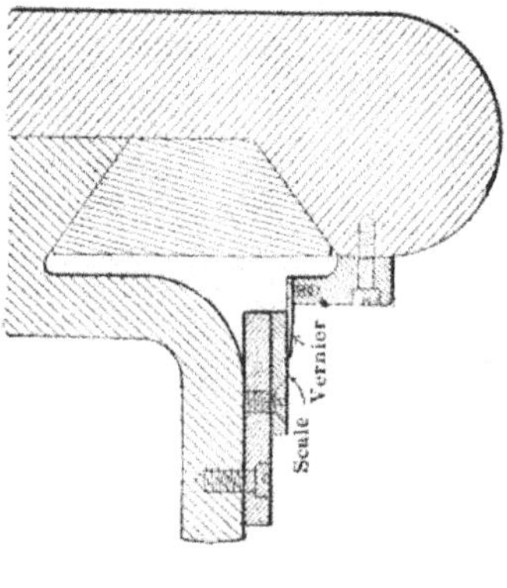

FIG. 53

MILLING THE JIG

In milling a jig of rectangular form that has finish on more than two sides, after measuring the casting to see that the finish is allowed for correctly, it is bolted down on the plate with the first face to be machined facing the spindle and milled with a face mill held in the spindle. After milling the first side, that side is turned down on the adapter and the work fastened for milling the second side, which will come square without the preliminaries of bolting it to a knee or otherwise squaring it up, which are necessary on a planer or shaper. In milling the third side, the work can be laid on the second side milled. To square it up, it is only necessary to insert a strip or two size blocks, which fit the slots, in the most convenient lengthways slot of the adapter and bring the jig up to them. To avoid cutting the adapter with the mill, the work should project over the edge of the adapter more than the amount of finish allowed and, if the blocks do not bring it that way, parallels should be inserted between the blocks and the jig to make up the difference. If either end is to be milled, the blocks should be inserted in one of the crosswise slots and the side of the jig brought up to them, thus bringing the end square, both with the side on the adapter and the one against the blocks.

INSIDE MILLING

After finishing the outside of the jig, the open side is turned up, the face mill removed from the spindle and the vertical fixture put on. With this the locating spots and any finished bosses there may be on the inside of the jig are rapidly and accurately milled. Where the tools are kept in good shape, the work does not peen as much as in using surfacing mills and does not seem to get out of shape from the cutting as much as in removing the scale by planing, so that it is rarely necessary to go around the work more than once unless the requirements are very close.

In milling flat plate jigs the vertical attachment is put on the first thing; after milling one side, the work is simply turned over to mill the other.

THE VERNIER ATTACHMENTS

In boring, the feed screws are ordinarily not accurate enough to use for very close measurements in moving the work to get the proper locations of the different holes, and without special means of measuring it is necessary to bore a trial hole undersize, make a plug to fit it and measure its position with micrometers or vernier calipers before finishing it. When verniers are attached to the machine, the proper location can be ascertained from them at once, and the holes finished at the first setting.

As stated in Chapter I, verniers and scales for use in connection with them are made by the Brown & Sharpe Manufacturing Companv. For an ordinary universal milling machine two scales are needed and two verniers, one scale as long as the longitudinal travel of the table and one as long as the vertical adjustment of the knee.

The scale for the table is attached to the front lower edge of the table and its vernier attached by a bracket to the saddle in the place ordinarily occupied by the feed-controlling mechanism, as in Fig. 52. The scale for vertical adjustment should be attached to the left side of the column close to the knee, so that the vernier can be bracketed to the knee, as shown in Fig. 53.

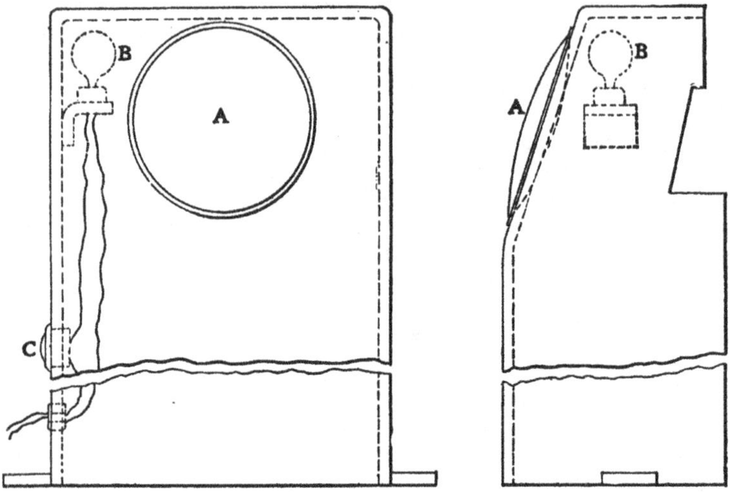

FIG. 54. — VERNIER BOX

A CONVENIENT ARRANGEMENT

To make things more convenient, in some shops the vernier is covered by a small metal box such as is shown in Fig. 54, in the top of which is set a French reading glass *A*. *B* is a small electric lamp set in one corner of the box in such a position that its light will shine on the vernier, but will not be thrown in the operator's eyes. By pressing the button *C* this lamp can be lighted at any time and greatly facilitates getting a correct reading.

Flat plate jigs cannot be worked on a machine equipped in this manner without putting them up on an angle plate, unless there are also a scale and vernier for the movement of the saddle upon the knee, in which case the vertical attachment can be used for boring. It is preferable, if the amount of work will warrant it, to have a vertical machine for this class of work, with verniers for the table and saddle movements only.

HANDY BORING TOOLS

Some styles of tools for use in boring are shown in Fig. 55, in which *a* is a spotting drill, *b* a chuck for small drills, *c* a flat drill for cast-iron work, and *d* and *e* types of boring tools suitable for boring single holes. Tool *d* is best for steel, as it is easy to set so that it will cut within 0.002 inch to 0.003 inch by measuring over the bar and the point of the cutter with a micrometer, the reading being the diameter of the bar + ½ (diameter of hole to be bored — diameter of bar). The hole can be sized with a reamer. It is not safe to use a reamer on cast iron; for that material *e* is a better boring tool on account of the refinement of adjustment which can be obtained when boring to a plug.

USING A BORING BAR

When a pair of holes are to be bored in line with each other in opposite sides of the jig, or when a single hole is to be bored some distance out from the spindle, it is necessary to use a bar long enough to reach through the jig plus the length of feed necessary. In this case the bar should be put in the machine before the work is put on the adapter, and the overhanging arm with a bushing fitting the bar be set so as to bring the bar square or parallel, as the case may be, with the locating strip on the adapter. Then without moving the arm, the bushing and bar can be removed and the jig set up for boring.

Of course on drawings of jigs to be machined in the manner described, the positions of stopping places and holes should be figured in two directions at right angles to each other from the sides or ends of the jigs.

STOPS FOR SETTING THE MILLER TABLE [1]

In connection with the handling of jig work on the miller, another method making use of the stops shown in Fig. 56 should be of interest.

As will be noticed, in the saddle of the machine there is a T-slot milled for a ¼-inch bolt; a similar but larger slot is shown in the table which usually carries the dogs for tripping the feed. Into these slots parts *A* and *B* are fitted, the essential parts being made in correct alinement with one another. *A* carries a 1-inch micrometer head, and *B* a nurled-head screw which is used to facilitate the setting of the micrometer at any required point. The screw, which is made with a fine thread, has a rounded point and acts as an anvil to the micrometer head. This arrangement is for setting the table where the holes are bored in the jig 1 inch or less apart and needs no explanation. Standard end-measuring rods can be used in getting lengths longer than 1 inch.

[1] Contributed to the *American Machinist* by C. I. Press.

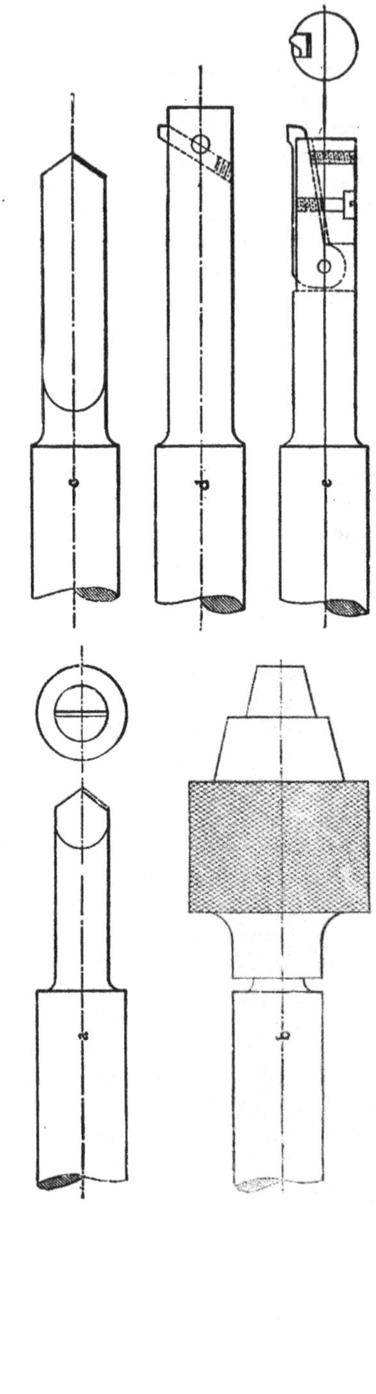

FIG. 55. — DRILLING AND BORING TOOLS

In the front side of A and B there are fitted and soft-soldered in place two hardened buttons C C; these are used in connection with a vernier for taking measurements over 1 inch in length, the vernier being in correct position when the sides of the jaws are held against the shouldered part of these buttons. The jaws of the vernier take the measurement over the

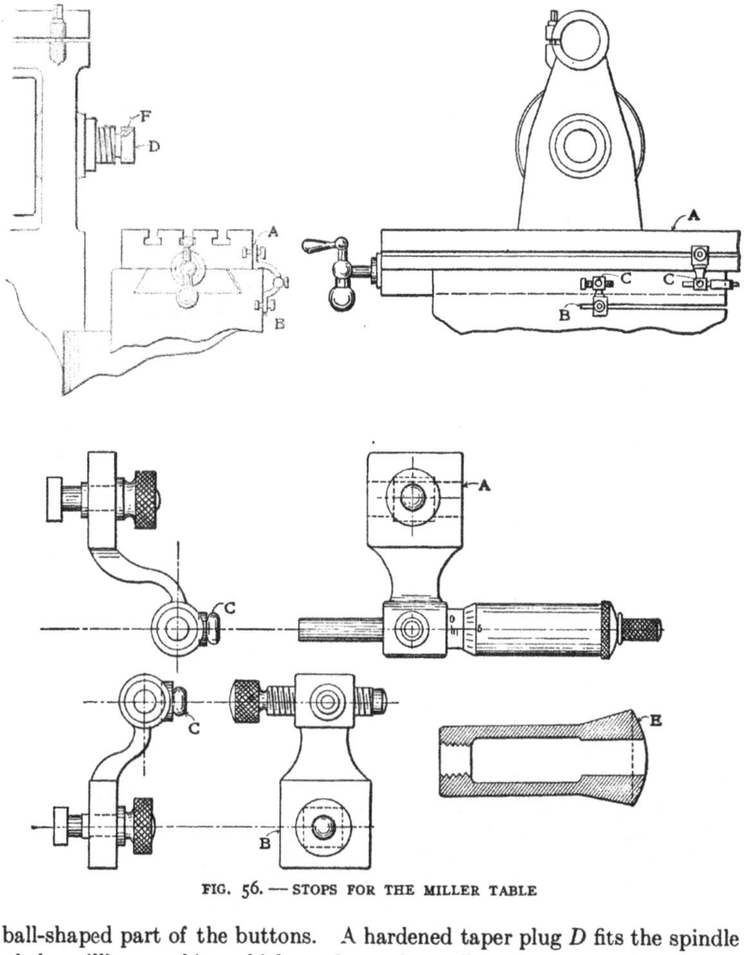

FIG. 56. — STOPS FOR THE MILLER TABLE

ball-shaped part of the buttons. A hardened taper plug D fits the spindle of the milling machine which carries spring collets, a section of one being shown at E. These collets need no explanation, except that they carry all tools needed in boring, drilling, and reaming operations. They are threaded internally at the rear for the draw-in rod which passes through the spindle of the milling machine and are clamped by a hand wheel in the usual way. A soft sleeve *F* is forced on plug *D* and becomes a permanent part of that fixture. Before starting on a piece of work, *D* is driven in place, and a very light smooth cut is taken over *F* with a tool held in the milling-machine vise. This forms a true place to take vertical measurements from by the aid of a hight gage used as shown in chapter I. It is a good idea in this connection to use a strip of tissue paper as a feeler between the jaw of the hight gage and *F*.

CHAPTER IV

BORING HOLES ON THE MILLER AND CHECKING WITH VERNIERS[1]

SEVERAL years ago in a certain shop a number of very fine perforating dies, also a number of drill jigs, were to be made. All the holes in these tools were to be accurate to within one-thousandth of an inch, and from past experiences it was felt that the graduations on the screw of the milling machine which was available, could not be trusted.

THE CHARACTER OF THE WORK

The tools to be bored were of such dimensions that it was necessary to clamp them up on an angle plate on the milling-machine platen in order to get sufficient travel of the platen in both directions to cover the holes to be bored. This made the problem of locating the holes accurately more difficult than it would have been had it been possible to clamp the tools down flat on the platen and use the vertical head of the miller to carry the boring tools. A moment's consideration will show why this statement is true. With the tool to be bored clamped upon edge, it was necessary to adjust the knee of the milling machine up and down a sufficient amount to take in both the lowermost and uppermost holes, and this adjustment of the knee resulted in throwing the work out of square. It would, of course, seem that, with the gibs of the milling machine nicely adjusted, it should be possible to raise and lower the knee without throwing it out of square; but the fact remains that when a tool of rather large area is to be bored, thereby necessitating a considerable movement of the knee as well as a considerable longitudinal movement of the platen, the work is likely to be thrown out of square owing to various conditions of overhang and slight errors in the milling-machine construction.

HOW THE PLATE WAS HELD

In considering this problem, it was decided that not only must some way be devised for accurately spacing the platen lengthwise, and also accurately spacing the knee up and down in order to locate the holes properly, but that also some method had to be provided for checking the holes in a manner to guard against overhang of the knee and platen. It

[1] Contributed to the *American Machinist* by J. G. Vincent.

was finally decided to use the following arrangement, and it worked out very satisfactorily:

Figs. 57 and 58 illustrate the manner in which the plate to be bored was clamped on the milling machine. It will be noted that a parallel strip is placed under the plate in order that the lowermost hole may be a sufficient distance from the platen to facilitate measuring. In securing the jig plate in this position, two angle plates were used, one at each end, in order that none of the holes to be bored should be covered. In order to facilitate the work, the holes were laid out roughly and drilled to within a sixteenth of the finished size before the jig plate was clamped on the milling machine, thereby making it unnecessary to remove the boring bar after the job was completed. It would be better practice, however, to rough drill the holes by using a drill in the spindle of the milling machine, locating the holes by means of the dials on the screws, thereby locating them quite accurately and economically and leaving very little stock to be taken out during the final boring operation. In this case the holes should be rough drilled to within a thirty-second of the finished size and still leave plenty of stock for finish boring.

THE SWEEP INDICATOR AND THE TEST PLUGS

In clamping the jig plate on the milling machine as just described, a sweep indicator was used in the milling-machine spindle to insure the location of the jig plate at right angles to the spindle, paper being used to shim up between the jig plate and the angle plates.

In order to illustrate clearly the principle involved in Figs. 57 and 58 an imaginary jig plate is shown in which eight holes are to be accurately located, according to figures. To simplify the explanation, it is assumed that all the holes are to be bored 0.75 inch in diameter; in view of this fact, it is only ncessary to have one special plug, such as is shown in the lower right-hand hole (Fig. 57). It is to be understood, however, that where the holes to be bored vary in size, it is necessary to have special plugs, the larger diameters of which correspond to the holes to be bored, while their smaller diameters are all of the same size, such smaller size corresponding in diameter to the diameter of the boring bar.

A boring bar of 0.500 inch diameter has been found a good size for all-around work, and the smaller ends of all the special plugs are therefore made 0.500 inch to correspond.

As shown in Fig. 57 it is necessary to have the length of the hypotenuse of the angles between the different holes. The solving of this should be done in the drafting room and a drawing of the jig plate made with dimensions along the lines shown in Fig. 57. For the jigs mentioned, however, there was no drawing, except one of the part the jig was to make, and the dimensions had to be solved as the work proceeded. This can readily

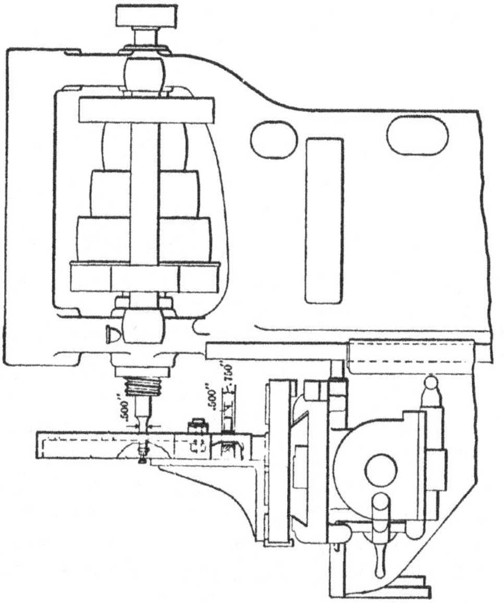

FIG. 58

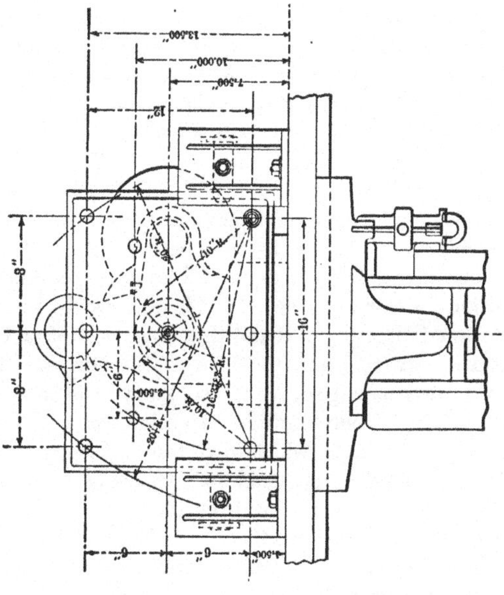

FIG. 57

LOCATING HOLES ACCURATELY IN SPECIAL TOOLS

be done if the toolmaker is fairly proficient in mathematics, but it is the best practice to do this work in the drawing-room, care being taken to give the dimensions in such manner that the toolmaker can work directly to such dimensions without complicated calculations.

BORING AND CHECKING THE LOWER HOLES

After the work had been set up as just described, the holes were bored as follows: First the lower right-hand hole was bored to the finish size, and then the special plug gage, as shown in Fig. 58, was inserted, it being understood, of course, that the cross feed of the miller was used to do the feeding. After this hole was bored, the platen was moved lengthwise to bring the work in position to bore the lower center hole. In making the final adjustments preparatory to boring this hole, the distance between this hole and the center one just finished was located with a vernier caliper by measuring from the plug and the boring bar, it being understood that in taking these measurements 0.500 inch was allowed for the two half diameters of the special plug and boring bar. The up and down distance was located with the hight gage, by measuring under the special plug with the hight gage, setting by the platen, and then measuring under the boring bar. In this case these two measurements should be the same.

After the lower center hole had been bored, the platen was again moved in position to bore the lower left-hand hole; in making final adjustments preparatory to boring this hole, the distance was checked in the same manner as the center hole, a long vernier caliper being used in order to make it possible to measure from the initial hole at the right and thereby avoid accumulated errors. After boring the lower left-hand hole, the work was next lowered to position for boring the upper left-hand hole.

HOW THE OTHER HOLES WERE LOCATED

By using a 6-inch block which had previously been scraped to a true cube, it was possible to check the distance between the platen and the upper left-hand hole with the hight gage. In lowering the knee to bring the work in position for this upper left-hand hole, it would seem that the work would be lowered directly down and that no change of platen lengthwise would be necessary to bring the hole in the correct longitudinal position; but from previous experience it was known the work could not be trusted to come right. Therefore this dimension was checked by solving the hypotenuse of the angle, which gave the dimension between the initial hole bored and the upper left-hand hole, which, as shown in Fig. 57, was 20 inches. By carefully measuring from the special plug to the boring bar, this dimension was checked, the platen being simply adjusted to bring the dimension correct. After this dimension had been checked, the vertical dimension was again checked to make sure this dimension had not been altered; then the hole was ready to be bored. It is apparent from the figures how all of the holes could be checked in the manner described. It will be noted, for instance, in boring the center hole of the plate, that not only was the hypotenuse checked from the initial hole, but also from the lower left-hand hole, as in this case both dimensions were the same.

ACCURACY OF THE PROCESS

In conclusion it may be added that this scheme of checking the work has been used on numerous complicated jigs and dies, and without boring a hole that was out to exceed one-thousandth of an inch. One perforated die bored in this manner contained fifty-six holes; after the die was completed, two plates perforated by it could be reversed and a plug inserted through all of the holes, thereby showing that the work was practically perfect.

CHAPTER V

A Precision Drilling and Reaming Machine[1]

The machine shown in Fig. 59 and the method described in this chapter are used by the Burroughs Adding Machine Company, of Detroit, Mich., in its experimental room. This method has been developed and adopted, after many years of experience, as the best and quickest means of getting out the different parts when some improvement makes it necessary to alter or redesign the machine.

LAYING OUT THE WORK

As in all well-regulated manufacturing plants, the ideas originating in the brains of its inventors are first worked out and put on paper by a corps of draftsmen.

In this case a vertical line is placed on the drawing paper, the center of the main shaft is located and another line drawn through this center at right angles to the vertical line. The location of the main shaft, a little above and to the left of the center, is shown in Fig. 60 by the largest hole in the plate. Each piece to be made is then drawn on a separate sheet of paper. All dimensions are given from the vertical or the horizontal line, or both.

This system of drawing makes it much easier for the draftsman to make an assembled drawing, as each piece is shown in the position it occupies in the machine. Each single-piece drawing can be laid on top of the assembled drawing and the part pricked through with a pin to see that it does not interfere with any other piece in the machine; or the piece on the single drawing can be transferred to the assembled drawing with less work than if drawn in any other position.

It also simplifies the work of the model-maker so that by moving the platen in, out or sideways to conform with the dimensions on the drawing, he can lay out and drill the piece on the specially arranged drill shown in Fig. 59.

The adding machine is shown by Fig. 61 and a few of the parts in Fig. 62. The side frame of the machine will be seen in the foreground of Fig. 60. The machine is composed of a large number of pieces, each working in unison with many others; hence each piece must be worked

[1] E. F. Lake in the *American Machinist*.

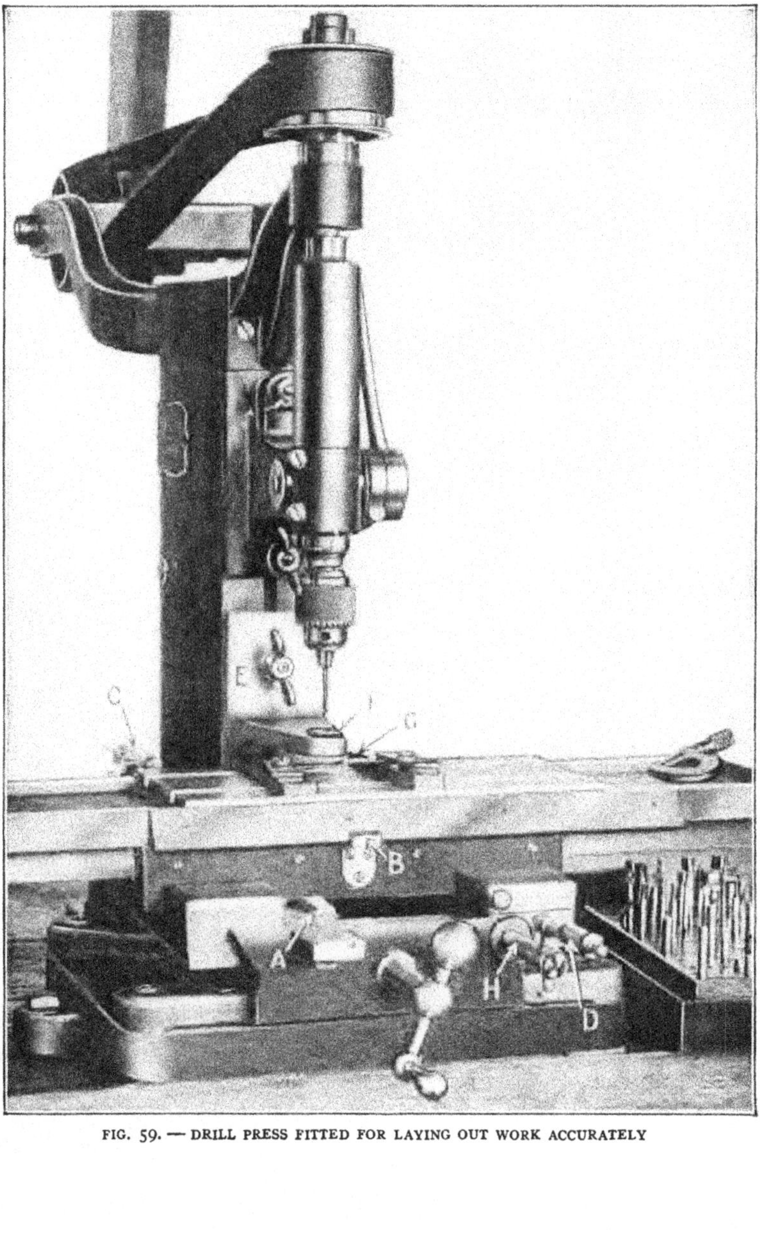

FIG. 59. — DRILL PRESS FITTED FOR LAYING OUT WORK ACCURATELY

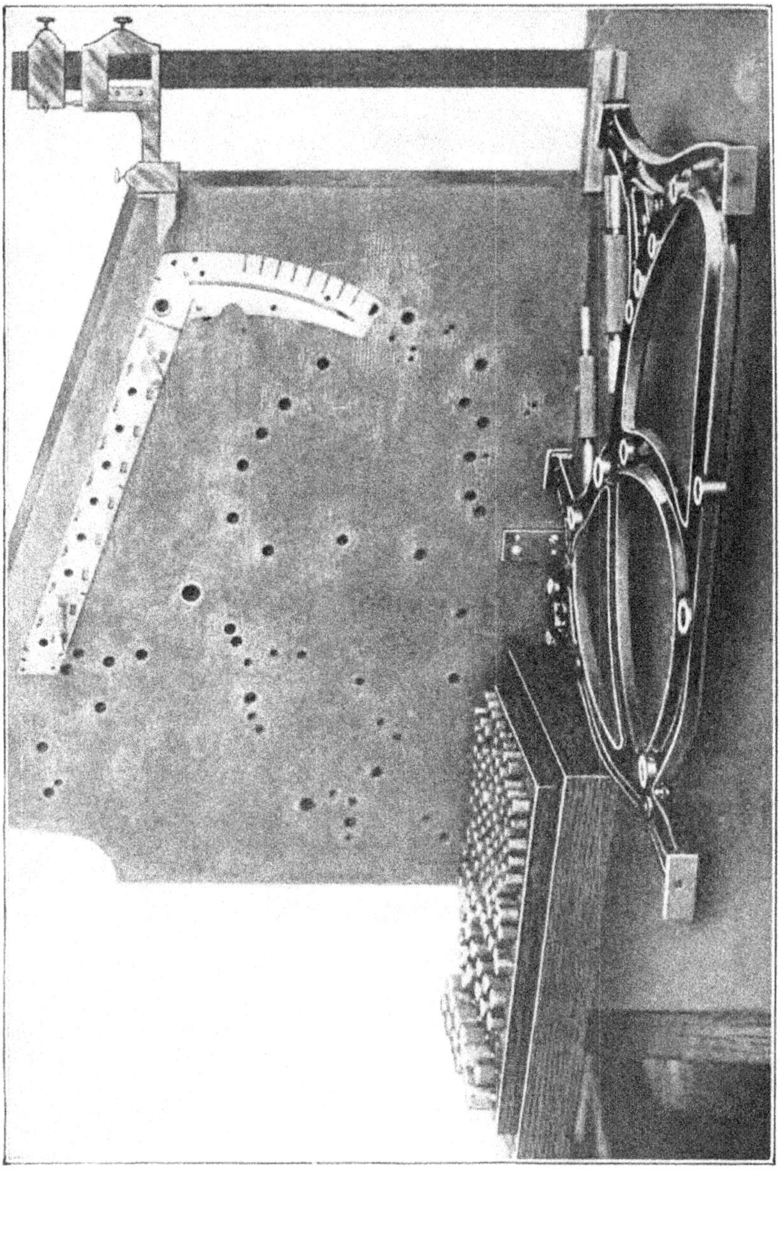

FIG. 60. — THE PLATE ON WHICH ALL PIECES ARE TRIED TO DETERMINE THAT EACH HOLE IS AN EXACT FIT

out to the exact sizes given on the drawings, as a variation of one-thousandth inch might throw some other part out so much that the machine would not work.

FIG. 61. — ONE STYLE OF THE COMPLETED BURROUGHS ADDING MACHINE WITHOUT GLASS CASE

THE DRILL PRESS

In laying out and drilling this work, an ordinary sensitive drill as made by the Sigourney Tool Company is fitted with two verniers, as shown at *A* and *B* in Fig. 59. One is used when the platen is moved in and out, and the other when it is moved to the right or left, so that adjustments can be made to a thousandth of an inch. At the back of the platen, as shown at *C*, is located a Boulet indicator. As this indicator has a scale laid out on an arc and graduated to thousandths of an inch, the platen can be moved out and returned to the exact position, after the work has been clamped to it. A pointer moving back and forth on the indicator gives the readings. This pointer in turn is moved by a plunger which presses against the platen when it has been returned to its original position.

At *D* is located a micrometer attachment with a stop by means of which the platen can be located, moved away for other operations and returned again to its original position. Where the distance between the centers of holes is but a few thousandths, by starting at the outside hole

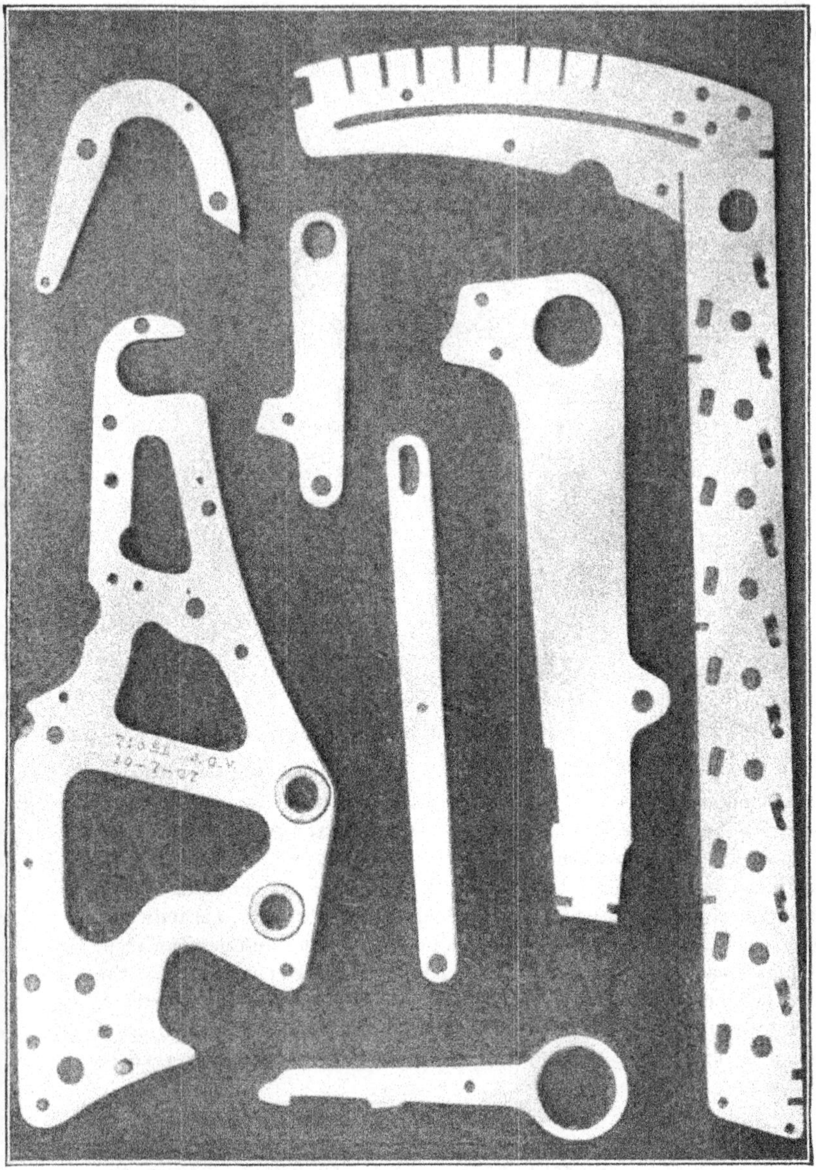

FIG. 2. A FEW OF THE PIECES LAID OUT ON THE DRILL PRESS

and drilling that first, the platen can be moved in by the micrometer and then moved to the right or left by the end screw, and the dimension located by the vernier at *B*. To give the micrometer a range of several inches, a shaft *H* is fastened to the bed of the drill press. The micrometer is moved in and out on this shaft and clamped to it by a thumb screw on the under side which presses a split bushing against the shaft.

At *E* is located a knee containing a hole to receive the bushings *F*. The hole is trued up with great care and the bushings ground to an exact size and fit so that the drills, reamers or counterbores used in the machine, and shown in the case to the right of the drill press, will not be thrown out of true.

The work is clamped to the machine as shown by the piece at *G*. A piece of sheet metal is taken and a start made by drilling the most important hole first; then the platen is moved to make all other holes correspond to the dimensions given from this one. The method is best illustrated by the following description.

WORKING A PIECE

To lay out and drill the piece shown by Fig. 63, a piece of sheet metal of the correct thickness and size is taken and a ½-inch hole drilled at *A*;

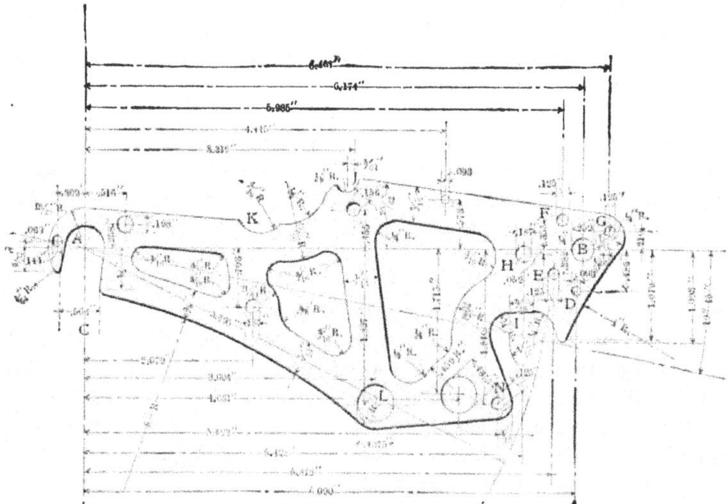

FIG. 63. — A PIECE AS LAID OUT AND DRILLED

this is then reamed out to 0.503 inch. The platen is moved to the left 6.174 inches, a ¼-inch hole drilled at *B* and reamed out to 0.252 inch.

The platen is then moved in, or toward the back of the drill press, 0.486 inch and to the right 0.84 inch, which makes it 6.090 inches from the perpendicular line *C*. The hole *D* is drilled and reamed to 0.093 inch in diameter. The platen is then moved to the right 0.278 inch, making it 5.812 inches from line *C*, and out enough to make it 0.282 inch from the vertical center line, where a $\frac{1}{8}$-inch hole is drilled at *E*. The same size hole is also drilled at *F* and *G* after moving the platen to correspond with the dimensions given in the drawing. The other holes in the piece are located, drilled and reamed in a like manner.

To get the opening at *I*, the platen is moved so that the centers from which the $\frac{1}{4}$-inch radii are struck will come under the center of the drill; then a $\frac{1}{2}$-inch hole is drilled at each place and the balance of the stock worked out by other means. The large irregular-shaped hole in the center of the piece is worked out by using the same $\frac{1}{2}$-inch drill to locate the three points marked $\frac{1}{4}$-inch *R*, and a $\frac{5}{8}$ drill to locate the point marked $\frac{5}{16}$-inch *R*. The other two irregular-shaped holes are worked out in the same manner; but as these are merely put there to lighten the piece, exactness is not required in locating the holes and no dimensions are given from the center lines.

The semicircular opening at *J* is obtained by moving the platen 3.319 inches less $\frac{5}{64}$ inch to the left of the center of hole *A*, and $\frac{25}{32}$ of an inch out from the same center. After drilling with a $\frac{1}{4}$-inch drill, the opening at *K* is located by moving the platen to the proper position and drilling a-$\frac{5}{8}$-inch hole.

TESTING

After the holes are all drilled, each piece is tested by locating it on the plate shown in Fig. 60. This plate is made of a cast-iron block in which are drilled holes that are bushed with hardened-steel bushings so that there will be no wear after inserting the plugs many times. The plate is also squared on the bottom so that it can be stood on edge on a surface plate. A vernier hight gage, as shown at the right of the half-tone, is used to check all hight dimensions.

In this plate are located the important holes of each piece of the adding machine, from the cast-iron side frame shown lying on the surface plate to the smallest sheet-metal lever, plate or bar. An idea of the number of the pieces which go into a complete machine can be obtained by a glance at Fig. 61, which will impress upon one the necessity of being exact with the dimensions of every piece.

The single-end plugs, which are a tight fit in the bushed holes in the plate, are shown in the case to the left of the plate. They are ground to an exact size so that no variation can occur in testing the pieces. A few double-ended plugs are used, similar to those shown on the face plate between the side frame and the plate.

The testing insures correctness in each piece before starting to assemble the machine and results in the assembling being done easily and quickly.

FINISHING THE PIECES

After drilling and testing the holes, the round corners scribed from the same center as a hole are obtained by inserting a plug with a head of the diameter that the round should be; on the other end of the plug is then placed a washer of the same diameter as the head. The plug and the washer have been ground and lapped to the exact radius given for the round corner. After they are placed in position, the metal can be filed away until it is the same size as the head and washer of the plug. For instance, to get the round at the left-hand end of the piece shown in Fig. 63, a plug would be made 0.503 inch diameter in the body and $\frac{15}{16}$ inch diameter on the head, with the washer the same diameter and a hole through it 0.503 inch in diameter. The plug is slipped through the hole *A*, the washer put on the end, the whole held in the vise and the end of the piece filed down to the head of the plug; as the plug is hardened, it will not allow any more filing, and the required radius of $\frac{15}{32}$ inch is obtained.

This operation is repeated with different-sized plugs at L, *N* and *G*, giving the correct lines to work the rest of the shape of the piece to at these points.

The rest of the form is worked out by the use of a filing machine. An ordinary bench filing machine is fastened to a bench plate together with a small electric motor which can be connected to any electric-lamp socket. This makes a very handy outfit, as it can easily be carried by the workman to any part of the shop in which he desires to do the work.

With the aid of a magnifying glass, and a little skill, the work can be brought to within less than a thousandth of the size desired.

IMPORTANCE OF CLOSE WORK

This method of laying out and drilling work shows the precision used by the company in building its first or experimental machine. The same careful workmanship is demanded when the tools, dies, jigs, etc., are made for manufacturing the machine in quantities, as the number of separate pieces in the smallest adding machines runs up into the hundreds. Each piece must be worked to the exact dimensions given, in order to work in conjunction or in unison with the numerous other parts it acts upon or comes in contact with.

In the completed machine as shown by Fig. 61, when a key is struck it not only prints the figure required but operates other parts which record the figure in the machine. When a total is desired, the pulling of a lever on the right-hand side of the machine, or the touching of a key which sets in motion an electric motor that operates the lever, causes numerous other parts to come into operation and print the total of all the figures previously registered.

As will be seen, this method of laying out will effect a great saving in time and enable the workman to work to the exact sizes in many other kinds of tool, experimental, or model work besides adding machines. With other attachments for raising and lowering the platen or head with precision to give exact depths, it might be used to good advantage in die-making where certain depths are required.

CHAPTER VI

Master Plates and How They are Made[1]

A master plate is a parallel metal plate containing a number of holes, each of which corresponds to the center of some hole, recess, projection or curve in the piece of work or article for which it is made. These holes do not serve as guides for drills, counterbores or other cutting tools, but are used for correctly locating the master plate on the face plate of a lathe or other machine tool, each one of the holes fitting nicely over a true-running plug in the center of the face plate or spindle. For convenience all the holes should, if possible, be of one diameter.

In addition to these locating holes there may be a number of holes in the plate for clamping screws and dowel pins for securing the work.

The piece to be operated on is fastened (usually soldered, or doweled and clamped) to the master plate, the whole then being strapped on the face plate, as shown in Fig. 64, and rotated or oscillated while the work is machined to correct measurements.

The position on the face plate of the work and master plate is changed from hole to hole as often as necessary to facilitate turning, boring, milling, grinding, etc., all of which operations (on small pieces) are usually done on the bench lathe.

MAKING A PAIR OF PLATES FROM A MASTER

Suppose, for illustration, we are manufacturing some machine or instrument which in a certain place requires two plates like *A* and *B*, Fig. 65, these plates being exactly alike as regards outline and position of holes, only one is recessed out on the opposite side from the other. In other words, they are right- and left-hand, or top and bottom plates. These plates are to be as near to correct measurements as possible.

The master plate for such a job is shown at *C*, Fig. 65. It has one hole for the center of each hole and curve in the work, or 12 holes altogether. The model blank is brought to shape by boring and, where it cannot be rotated, by milling with the bench-lathe milling attachment or by planing with a tool in the tool post or the slide rest.

It will be seen that if the holes used as centers for machining the

[1] Contributed to the *American Machinist* by E. M. King.

piece were inaccurately positioned and the work made to correct drawing dimensions for two or more curves, these curves would not meet or blend together. As the plate *B* will in its turn have to be fixed on the other side of the master plate from that to which plate *A* was soldered, it is plain that it will not be correct and match *A* unless all holes in the master plate go through it perfectly square and measure up exactly the same on

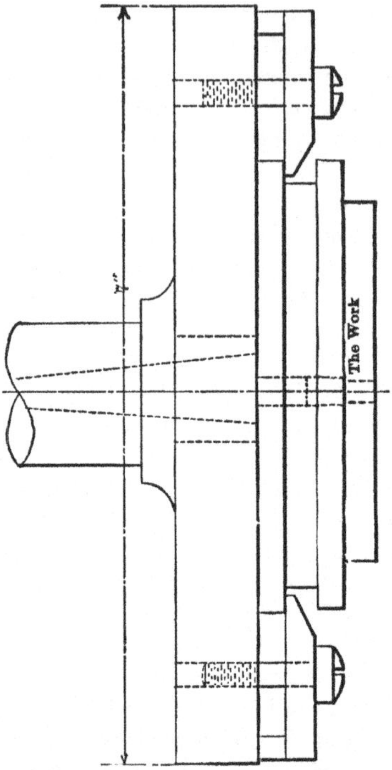

FIG. 64.—MASTER PLATE STRAPPED ON FACE PLATE OF BENCH LATHE OR QUILL

both sides. If the model is satisfactory, the master plate may now be used for making the blanking, piercing and shaving dies, also all necessary jigs, fixtures, depthing plates, etc.; or if desirable, it may be used for making the dies or molds for castings made by the die-casting process, the tools required of course depending on the material used in the plates and the method of manufacture.

In some cases the model (a watch for instance) is first perfected and then the master plate made by the model.

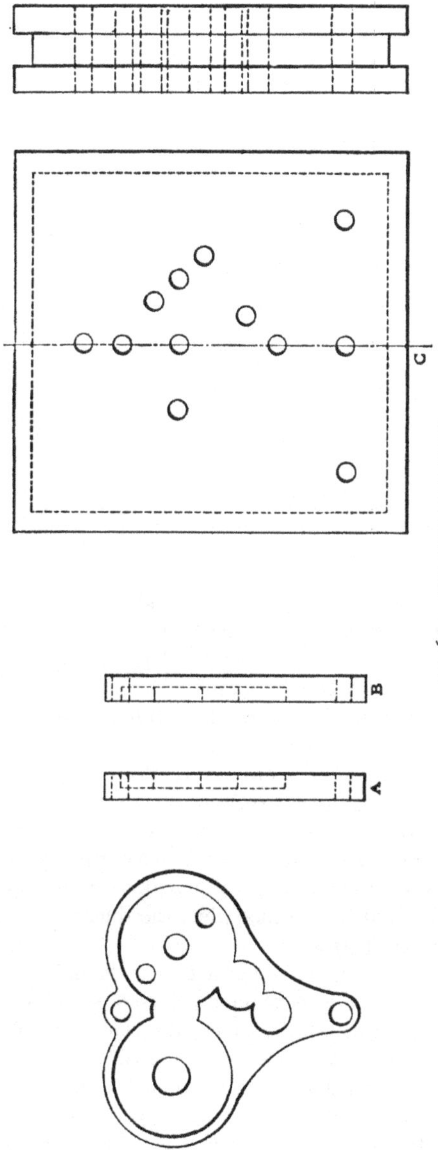

FIG. 65. — MASTER PLATE WORK

A GEAR-TRAIN PLATE

Fig. 66 shows a master plate for gear-train plates and connecting parts. The holes *a, b, c, d, e, f,* are for gear and pinion pivots, the others for dowels and screws. The limit of accuracy required is less than 0.0001 inch. From the foregoing it will be understood that the master plate, being the standard by which other tools are made, must be of more than ordinary accuracy. We will now consider the making of such accurate master plates.

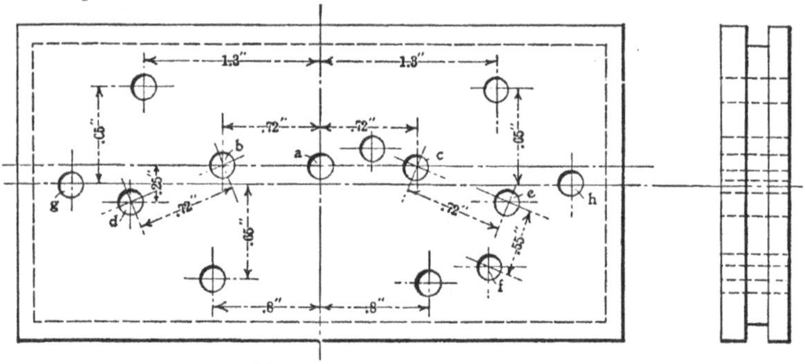

FIG. 66. — MASTER PLATE FOR A GEAR TRAIN

In the first place the plate must be made perfectly flat, smooth and parallel on two surfaces. This is accomplished by grinding on a good surface-grinding machine and afterward lapping slightly on a cast-iron lap which has been made true by scraping to a surface plate (or originated by the well-known method of scraping up three plates together) and then charged with fine carborundum or emery.

It is customary in some shops before the final grinding of tool-steel plates for master plates to heat them until the surface comes to a blue, it being believed that this relieves all strains and makes the plate less likely to change its shape. For some classes of master plates, Brown & Sharpe ground flat tool-steel stock is used without further grinding of the surface. This comes in strips up to 4 inches wide, the thickest stock, ¼ inch, being generally used for the plates.

There are several methods used for boring the holes in a master plate. If the shop in which the work is done possesses a precision dividing head or engine fitting the bench lathe, and if the work is small and a majority of the holes are in circles, they will probably be bored on the face plate of this head, using the milling attachment and a rotary boring tool. While this method is all right, in some cases it is rather uncertain, especially when the holes are not in circles. The method first described here is the button method, it being suitable for a large variety of work.

BENCH-LATHE EQUIPMENT

Suppose we wish to bore the plate shown in Fig. 66, it being already lapped true as described. For this work we will need a bench lathe in first-class condition, with a face plate which is perfectly flat and smooth and which runs in a true plane. A better plan would be to use a face-plate quill and quill rest, as shown in Fig. 67, which illustrates the quill rest, face-plate quill and driver of the Pratt & Whitney bench lathe. Owing to the design of these quills, they can be very accurately constructed, and, as the drive is taken from the lathe spindle by means of the quill driver which enters the spindle like a chuck while its open end slips over the clamp dog on the rear of the quill, there is no strain nor jar due to the action of the belt, to affect the accuracy of the quill.

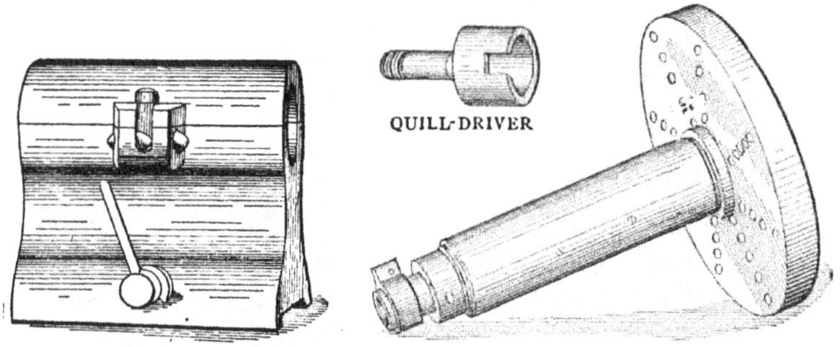

QUILL-REST FACE-PLATE QUILL

FIG. 67. — FACE-PLATE QUILL EQUIPMENT FOR BENCH LATHE

The face plate is solidly and permanently attached to the quill spindle which has a small taper hole in the end for the insertion of taper pins which can be ground true on the projecting end for the purpose shown in Fig. 64. The face plate must be ground true. The quill with the face plate is taken from the quill rest, the face lapped smooth by just passing it over the lapping plate, and the quill returned to place.

As the master plate shown in Fig. 66 is a little larger than the ordinary 7-inch bench-lathe swing, it will be necessary to have raising blocks for the headstock, slide rest and also for the quill rest, if we use one.

AN INDICATOR AND OTHER TOOLS

Almost all the tools necessary for this job are shown in the half-tone, Fig. 68. These consist of, first of all, a sensitive test indicator *A* showing very distinctly a movement of 0.0001 inch (this indicator multiplies 170 times); two or three buttons *B*, say 0.250 inch diameter and 0.3 inch high (though a 0.5 inch button is preferable where possible to use it); three or

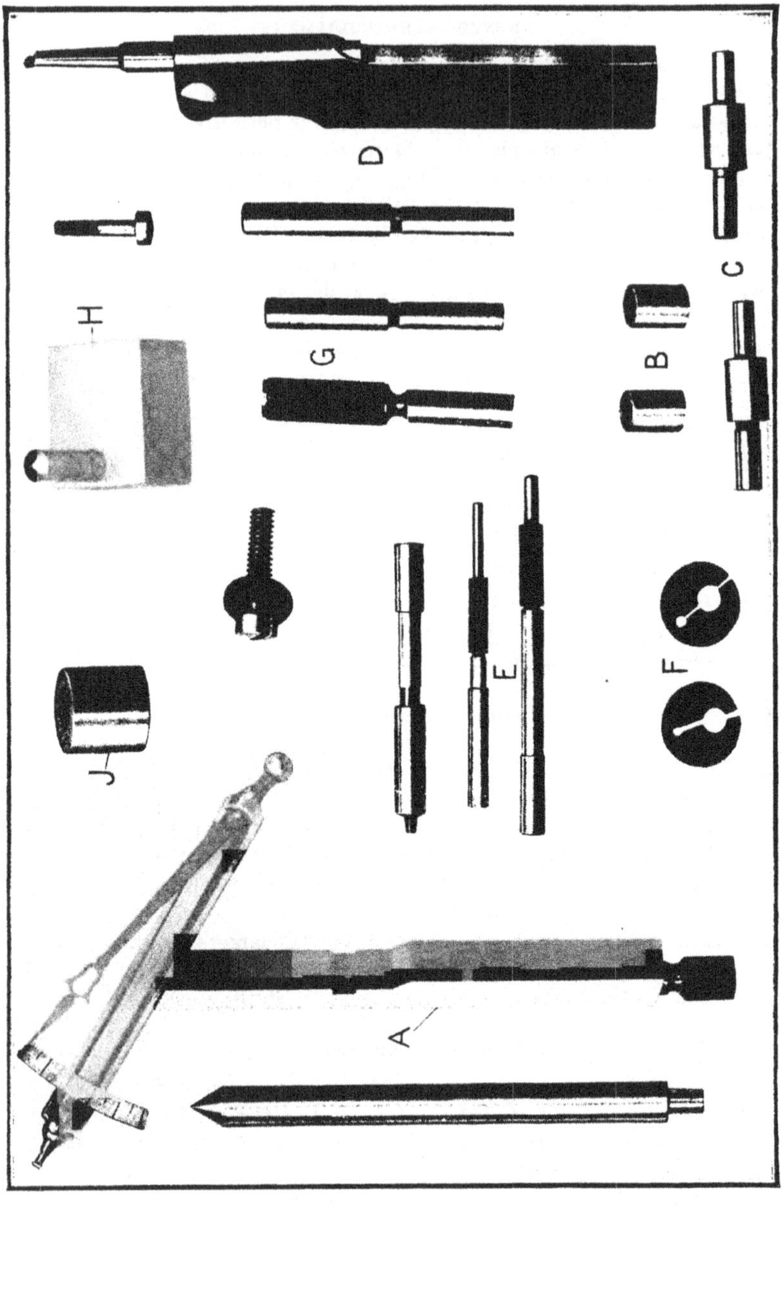

FIG. 68. — A SET OF TOOLS FOR MASTER-PLATE WORK

more plugs *C* of two different diameters each, the large or body size corresponding in diameter and length to the buttons, the small size fitting the holes in the master plate (in this case 0.160 inch); a boring-tool holder and tools *D;* a reamer not shown; some internal laps *E;* external laps *F;* plug gages *G;* a simple hight gage *H;* and one or two straight-edges, not shown. Button *J* is similar to *B*, only larger.

The buttons and plugs are hardened, ground and lapped. The buttons and large diameter of plugs are lapped to exact standard size. A good way to do this is to take a standard 0.250 plug gage which we are sure is correct and then lap the buttons and plugs until we can detect no difference between them and the standard when they are at the same temperature, using the micrometer as a gage with the spindle locked. The buttons are ground square on the bottom at the same setting when the diameter is ground. The shank on the right-hand end of the plug *C* is for the purpose of holding the plug in the chuck while grinding both diameters and the shoulder which rests against the master plate; it is also convenient for removing the plugs when the work is on the face plate. The reamer may be 0.0001 or 0.0002 smaller than the finished diameter of the holes. Two of the internal laps shown are made of copper and one of lead. Such laps are also easily made from round rods of soft solder by drilling a small hole and driving in a taper arbor having one or more flat sides, the outside then being turned off. The external laps *F* are copper. The gages shown at *G* are of slightly different diameters, the correct one being 0.160 inch.

LOCATING THE WORK ON THE FACE PLATE

Having these tools ready, we proceed with our master plate, laying it out by means of a surface gage or a vernier hight gage with scribing attachment; where the holes are to be bored, smaller holes are drilled and tapped part way through to fit the button-clamping screws. Now one of the buttons *B* is secured by a screw over hole *a* in the master plate, Fig. 66, and located with the vernier hight gage, the simple hight gage shown at *H*, Fig. 68, or by both together. The master plate with the button attached is then strapped loosely on the face plate with the button approximately central with the spindle. The indicator is clamped in the tool post, the contact point being about on the center as regards hight and touching the button either at the back or front. The lathe is rotated and the master plate rapped until it is in such a position that the indicator pointer shows no movement — or almost no movement. As the indicator multiplies 170 times, 0.0001-inch movement of the button would move the pointer over the scale 0.017 inch; or 0.0001-inch eccentricity of the button would show a pointer movement of 0.034 inch. If we bring the pointer movement down to 0.004 inch, this means that the button is less than 0.000012 inch out of correct position.

The surfaces of both master and face plates being true and smooth, strapping down tightly will not make any change in position. It is advisable to try the indicator against the surface of the master plate to make sure that it revolves in a true plane. After removing the button, drill, bore and ream the hole, paying particular attention to the boring and leaving but very little for the reamer to cut; in fact, the latter is run through merely to scrape the hole. Next take off the master plate and lap the hole, turning the lap by hand with a small handle or dog until the hole is a good fit for the plug gage.

BORING THE SECOND HOLE

The second hole *b*, Fig. 66, is next to be bored out. In order to locate the button for this hole, a plug *C*, Fig. 68, with small end 0.160 inch diameter and body exactly 0.250 inch, is inserted in the first hole *a*, and a button then placed over the hole already tapped at *b* and attached by a screw and washer. This button is located on the same horizontal line as the plug placed in hole *a*, by placing the master plate on one edge on the surface plate and sliding the hight gage shown at *H*, Fig. 68, under and touching both plug and button. This hight gage, it will be noticed, is merely a square block carrying a fine-thread spherical end screw which may be readily adjusted. The correct distance from the plug in hole *a* is found by measuring over both plug and button with a 1-inch micrometer caliper, of course deducting 0.250 inch from the measurement. This is the important measurement and no pains should be spared to get it

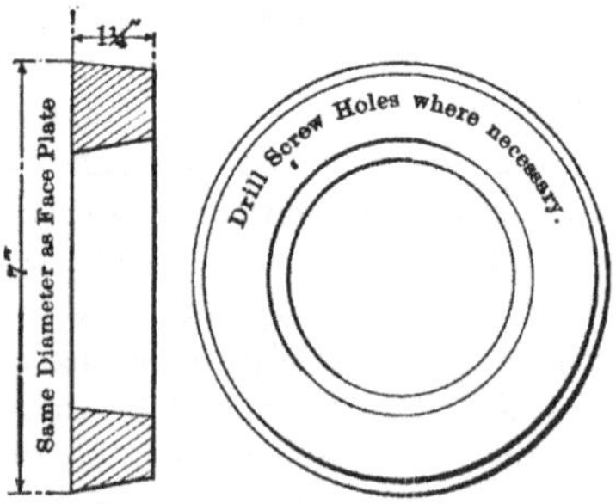

FIG. 69. — LEAD RING FOR COUNTER-BALANCE SEGMENTS

correctly. The master plate is now placed again on the face plate and the button trued up. The work is balanced conveniently by segments cut from lead rings like the one shown in Fig. 69, holes being drilled for the attaching screws. After trying the indicator again on the button, also on the surface of the plate, the hole is bored, reamed and lapped the same as the first hole *a*. When plugs are inserted in the two holes *a* and *b* and the distance measured, there should be no measurable error.

LOCATING AND BORING THE THIRD HOLE

The button for the third hole *c*, Fig. 66, is located in a similar manner to the one for the second hole *b*; but for final adjustment a good method is to place plugs in holes *a* and *b* and use a straight-edge on the three cylinders (the two plugs and the buttons), as shown in Fig. 70, moving it up and down as indicated by the dotted lines and noting on which cylinder it fulcrums, then trying it on the other side of the button and plugs. A still better way is to have two straight-edges, one on each side of the plugs. Then it will be instantly detected if both straight-edges do not act exactly alike.

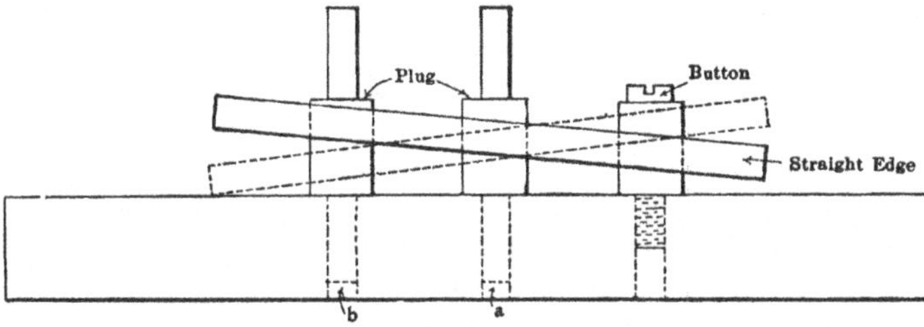

FIG. 70. — SHOWING METHOD OF USING STRAIGHT-EDGE ON BUTTONS

As we cannot use the ordinary 1-inch micrometer for measuring the distance from *a* to *b* or *a* to *c* when the plugs and buttons are in position, we must use something else. A Brown & Sharpe 3-inch or pocket vernier caliper is a very convenient tool for such work. In this case we use it simply as a snap gage, setting it to the outside measurement over the plugs in holes *a* and *b*, then rapping the button for hole *c* until, when gaged with the vernier thus set, the touch is exactly the same as for measurement *a b*.

A 2-inch micrometer may also be used to measure the distance from *b* to *c*. When the third hole *c* is bored, we should have these three holes exactly in a straight line, hole *a* being exactly central between holes *b* and *c*.

If the outside of the master plate from which we have been working is not now exactly parallel with the line through the centers of holes *a*, *b* and *c*, it can easily be made so by lapping.

FURTHER OPERATIONS

We next set buttons for holes *d* and *e*, Fig. 66, using the comparative method as much as possible, that is, making distances *b* to *d* and *c* to *e* equal *a* to *b* and *a* to *c*. We can also figure out distances *a* to *d* or *a* to *e* and make them equal, using a 2-inch micrometer as a gage. The hight gage *H*, Fig. 68, can be used for setting buttons for *d* and *e* the correct distance from the center line through holes *a*, *b*, *c*, without altering the adjustment of the gage from the former setting for the button for hole *b*, by putting two accurate 0.250-inch parallel blocks or gages under the edge of the master plate; or if the other edge of the master plate (the top edge in the illustration) has been used to work from, a 0.250-inch parallel block can be put under the hight gage, both being slid together under the buttons.

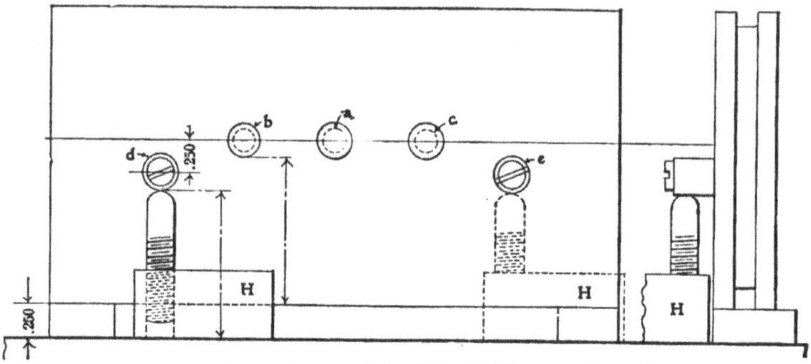

FIG. 71. — SETTING THE BUTTONS ON A MASTER PLATE

The scheme with the pair of 0.250-inch size blocks under the master plate will be clear from Fig. 71, which shows the hight gage *H* as used for setting the button for hole *d*, the adjustment of the gage remaining unchanged from its original setting for locating the button for hole *b*.

Another way is to use one of the straight-edges in contact with plugs in holes *a*, *b* and *c*, feeling with a 0.250 gage between the straight-edge and buttons *d* and *e*. Finally the distance from *d* to *e* may be tested with a 3-inch micrometer. Holes *f*, *g*, *h*, Fig. 66, may be located in a similar manner. If careful work is done as outlined in locating and boring, the holes will test up equally well on either side of the plate.

A MASTER INDEX PLATE

From the fact that, with the button method, comparative measurements of any number of holes or distances are easily made, it is particularly suitable for index plates and similar work. Fig. 72 shows a plate having nine holes, one in central position and eight others in a circle. The plate may be used for indexing directly or as a master for grinding the notches in other plates which are perhaps hardened; or we may have a hardened rim permanently attached to the original plates, using the holes for indexing during grinding only, or in case of repairs, or for testing.

The holes 1, 2 and 3 are located the same way as the three holes *a*, *b*, *c* already described in connection with Fig. 66. Hole 4 is located an equal distance from holes 2 and 3, the distance from 1 to 4 being, of course, the same as from 1 to 2 and from 1 to 3. Hole 5 is similarly located to hole 4, and preferably at the same time. The other holes are also located by comparison of distances one with another. With one gage, micrometer, or vernier caliper once set to gage the radial distances 1 to 2, 1 to 3, etc., another set to gage the chordal distances, and a good smooth plate, it is possible to set the buttons for the holes without a measurable error, even on quite large work.

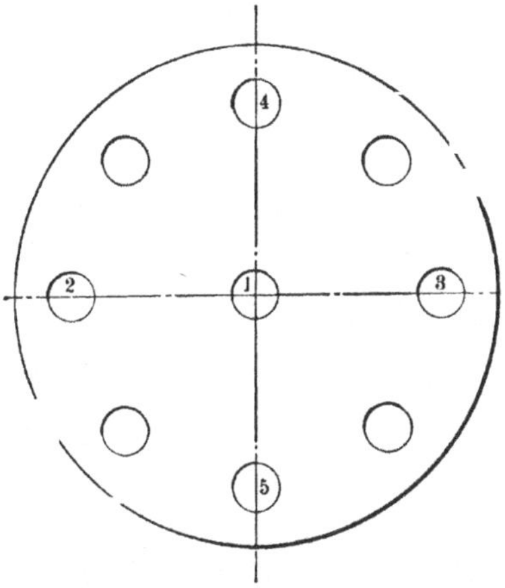

FIG. 72. — A MASTER INDEX PLATE

WORK ON THE MILLING MACHINE

For larger work than can be handled on the bench lathe, even by using raising blocks, a universal or plain horizontal-spindle milling machine fitted with a large face plate on the spindle, as shown in Fig. 73, can be used. This face plate should contain numerous tapped holes for use in strapping on the work, instead of the six or eight slots usually seen in engine-lathe face plates. The vise of such a machine forms an excellent tool holder. This arrangement is preferable to a large engine lathe with a rocker adjustment under the tool and no graduated dial on the cross-slide screw.

For heavy pieces or work which it is undesirable or inconvenient to rotate, accurate boring can be done on a good vertical-spindle milling machine, using the indicator and attachment shown in the half-tone, Fig. 74. The work is clamped on the horizontal platen so that the desired button can be brought under the spindle. If the machine is provided with spring or collet chucks, as several makes of machines are, grip the indicator in the chuck and rotate the spindle by hand, adjusting the indicator until it touches the button; then move the machine platen and work until the button is indicated exactly in line with the spindle. As in this case the indicator revolves round the button and its pointer cannot at all times be seen from the same view point, it may seem as though it would be very difficult to set the button accurately; but this incon-

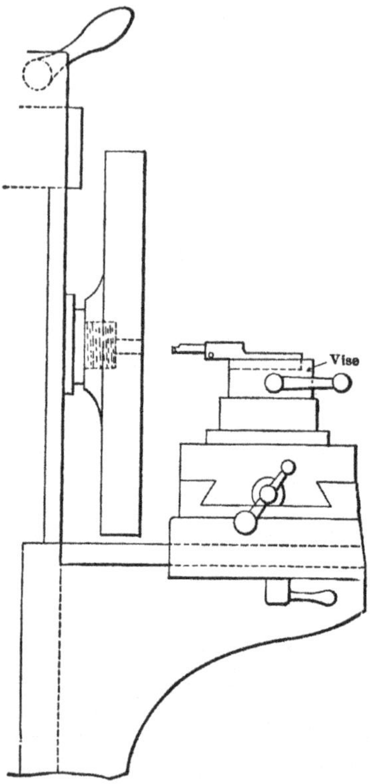

FIG. 73. — USING A FACE PLATE ON THE MILLER

venience is quite compensated for by the ease with which we can move the button and the work to position by the feed screws of the machine without employing the rapping method used on the lathe. Then we have no balancing to bother with; and if a number of buttons are already in position on the work, there is little danger of displacing one. A small mirror may be used in cases where it is otherwise difficult to see the indicator.

For large pieces it is well to have all the holes first drilled approximately correct, leaving only boring to be done on the miller. In such a case the buttons must be of sufficiently large diameter to cover the holes; they may be held on by screws with nuts which fit in small counterbored recesses in the under side of the piece to be bored. When the buttons are removed, these nuts drop into the T-slot in the platen of the miller.

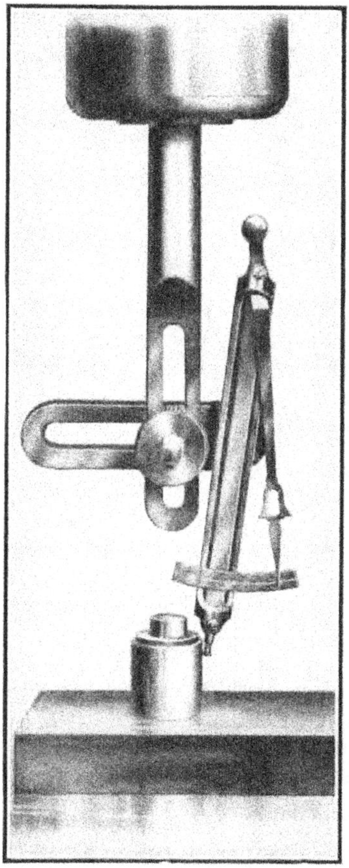

FIG. 74. — USING THE INDICATOR IN THE VERTICAL SPINDLE MILLER

THE USE OF DISKS ON THE FACE PLATE

An entirely different method from those described, of spacing holes in master plates is shown in Fig. 75. Take the very simple plate shown at *a* in the four sketches *A*, *B*, *C*, *D*, which plate contains six holes in all, three in each row, 0.5 inch apart, the two rows being also 0.5 inch apart.

The first hole is bored in the usual way on the lathe face plate as at *A*. The piece *b*, shaped like a try square and which has its two inside edges exactly at right angles, is now clamped about the master plate but separated from it by disks of suitable diameter. In this case the disks *c c* at the lower edge of the plate may be, say 0.2 inch diameter. The disk *d* to the right of the plate must be more than 1 inch diameter, say 1.2 inch.

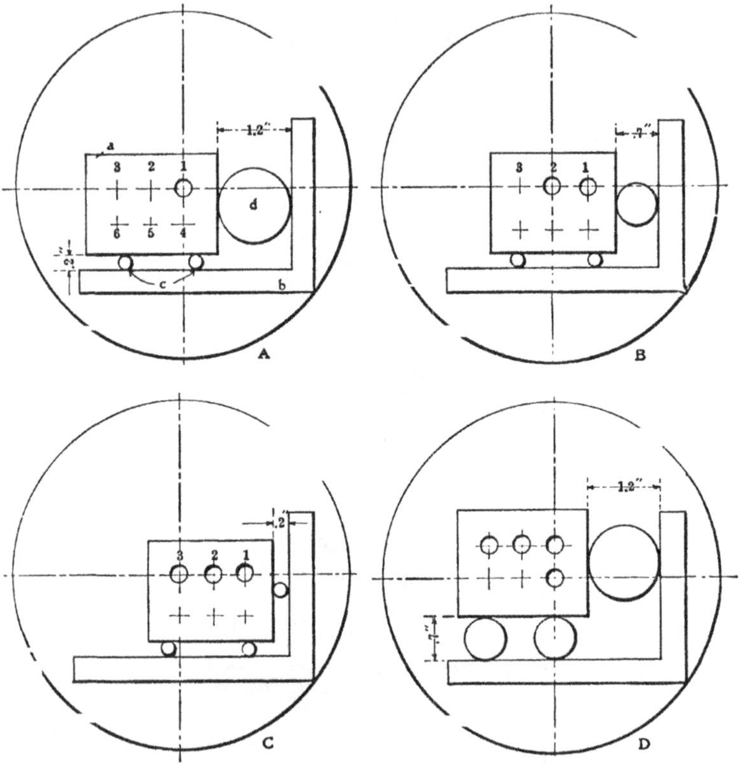

FIG. 75. — THE USE OF DISKS FOR SETTING MASTER PLATES FOR BORING

The straps holding the master plate are now loosened and the plate moved forward until it comes in contact with a disk 0.7 inch diameter, that is the plate is moved forward 0.5 inch in a straight line and in this position is shown at *B*. The master plate is now clamped to the face plate and hole 2 bored. Then the plate is again loosened and moved forward another 0.5 inch or until it touches a disk 0.2 inch diameter as at *C*. When hole 3 is bored the disks *c c* are removed and replaced by others 0.7 inch diameter, thus raising the master plate 0.5 inch as indicated at *D*. The operations described are repeated for holes 4, 5 and 6.

For this simple job we will need seven accurately ground disks, three 0.2 inch diameter, three 0.7 inch diameter and one 1.2 inch diameter. Instead of disks small parallel blocks can be employed, but disks being much easier ground on the bench lathe are generally used.

This method appeals to many people as being simple, quick and accurate. It certainly is a very quick method after the disks have been made and possesses the advantage that should one hole be spoiled or made too large all the holes can readily be rebored. Sometimes a very small amount can be rebored from all the holes without altering the adjustment of the boring tool.

INDICATOR DETAILS

The detailed sketch of the indicator shown in Fig. 76 may be of value in connection with the preceding matter relating to master plates as it is practically the same as the one illustrated in the group of tools in Fig. 68. It was designed principally for truing up buttons and small holes in face-plate work and for general use on the bench lathe. It may, however, be conveniently used for almost any kind of machine or bench work for which any indicator can be used.

The body or frame *A* is made from bright drill rod 0.277 inch diameter (J size). The lower lever *B* is not balanced. The pointer *C* is balanced with a $\frac{3}{16}$-inch commercial-steel ball annealed and sawed for the insertion of the pointer.

The shanks of the contact buttons *D* are made taper to facilitate changing for different styles of buttons, that shown being convenient for round work, but one about 0.3 or 0.5 inch diameter is sometimes desirable.

The nurled nut *E* is locked firmly to piece *F* (to which is attached the spring wire *H*) by means of check screw *G*, the whole being a friction fit in *A*, and is used for obtaining the spring pressure necessary to hold the pointer at either extremity of the scale *J*, according to which side of contact button *D* we wish to use against the piece being tested.

The scale *J* is a friction fit in *A*. The dimensions given on the drawing give the pointer a movement 171 times that of the contact point, but this can be easily modified by altering the distance of pin *K* from the pivots on which the pointer moves. The dimension from the pivot to the ball (0.22 inch) is approximate only, the correct distance for balancing the pointer being found by trial.

The construction of the spring center shown to the left of the indicator in Fig. 68 hardly needs any description. It is used in connection with the indicator for truing up work to center punch holes, on the lathe face

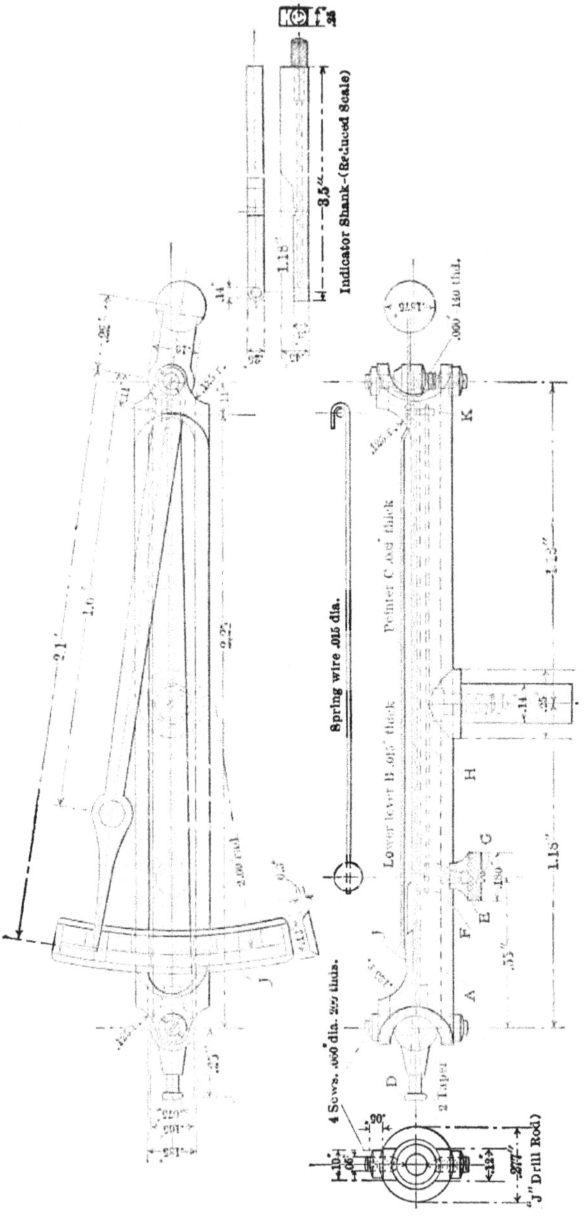

FIG. 76. — TEST INDICATOR (SHOWN ABOUT 1½ TIMES ACTUAL SIZE)

plate, it being possible to position such center marks accurately in about one-fourth the time required to set them even approximately with some of the old-style center indicators.

The attachment shown in Fig. 74 might be improved upon by off-setting to the left the part protruding from the chuck, so that the center line of the indicator could coincide with the center line of the machine spindle. It would then be more convenient for truing small holes in which no plug is used.

For attaching to a surface gage a retort-stand clamp may be used.

The drawings from which both this engraving and the indicator were made were furnished by S. Gahan, who, however, does not claim to be the originator of this style of indicator, but states that its present design is the result of suggestions received from many able mechanics, particularly W. H. Nichols, of Waltham, Mass.

CHAPTER VII

MASTER PLATES AND THEIR USES IN DIE MAKING[1]

As stated in the preceding chapter a master plate is a metal plate containing a number of holes having the positions of their centers corresponding with centers that have to be accurately located in a piece of work. The master plate is fastened to the lathe face plate with clamps, but the work may be fastened to the master plate with clamps or with screws by having the screws pass through the work and into the master plate. Or, very often, the work is soldered to the master plate, but soldering is only resorted to when it is not convenient to use clamps or screws. Soldering is sometimes dangerous for the reason that shrinkage of the solder on the side of the work is quite likely to draw the work sidewise, and thus the accurate location of the work with relation to the holes in the master plate is lost. Master plates are generally of steel, though occasionally for hurry-up jobs they are made of brass; if they are to be used indefinitely and for accurate work, they should be made of steel.

The accompanying half-tones and sketches are illustrations of actual shop practice at the works of the Veeder Manufacturing Company, Hartford, Conn., who are specialists in the manufacture of accurate machine-made castings which are made in steel dies operated by automatic machines. The use of master plates is invaluable in the construction of these dies and they constitute the only practical way to obtain good results when extreme accuracy is required.

A TYPICAL PLATE

In order to describe some of the methods used for producing master plates let us take as an example plate *A*, Fig. 77. This plate is made from machinery steel which with careful use gives a hole having a surface sufficiently hard to eliminate the necessity of hardened bushings. The plate is carefully ground to bring its two sides perfectly parallel. The sides of the grooves in the periphery are intended for clamping surfaces, and must be perfectly parallel with the outside surfaces of the plate as illustrated diagrammatically in Figs. 78 and 79. The correct way is shown in Fig. 78 and the incorrect way in Fig. 79. If not made as shown in Fig. 78 there will be trouble when attempting to move the master plate

[1] Contributed to the *American Machinist* by E. H. Crosby.

slightly sidewise for the purpose of truing a button or plug perfectly concentric with the axis of the lathe. It can readily be seen that there will be difficulty in controlling the sidewise movement when the piece is held on the face plate with three or four clamps which rest on an angular surface as in Fig. 79.

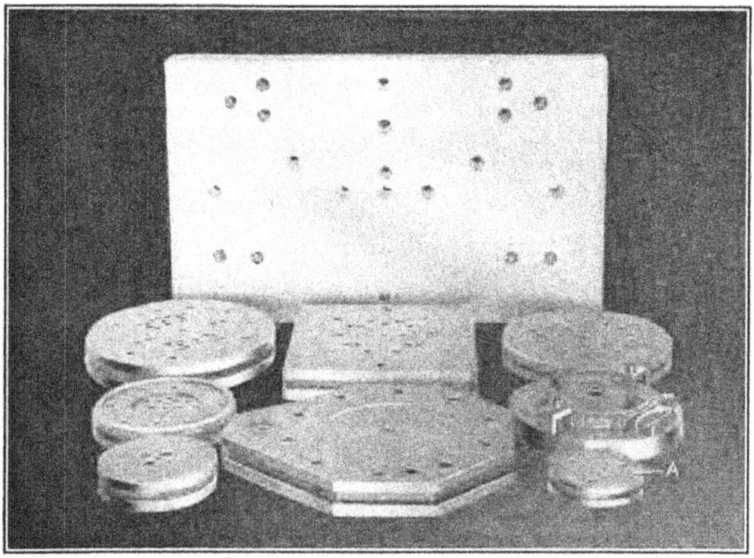

FIG. 77. — GROUP OF MASTER PLATES FOR DIE WORK

LOCATING AND BORING HOLES

In locating the holes *A*, *B*, *C*, *D*, *E*, *F*, *G*, *H* and *I*, which have a common diameter of 0.070 inch as shown in Fig. 80, we begin with center hole *A* by placing the master plate on the lathe face plate perfectly concentric with the axis of the spindle. We then drill a hole about 0.005 inch smaller than the finished size. It is then bored to within about 0.0002 of its correct diameter, then reamed, the reamer really acting as a burnisher rather than a cutting tool. This hole, together with the others, is first drilled and bored through one side of the plate. After placing all the holes in one side it is then reversed and each hole in turn placed on the center plug as in Fig. 81 and the hole bored to meet the hole on the plug. This method is used for boring the holes through the plate as they are likely to run and the inner ends of the holes be slightly out of true when they are carried all the way through the plate from one side.

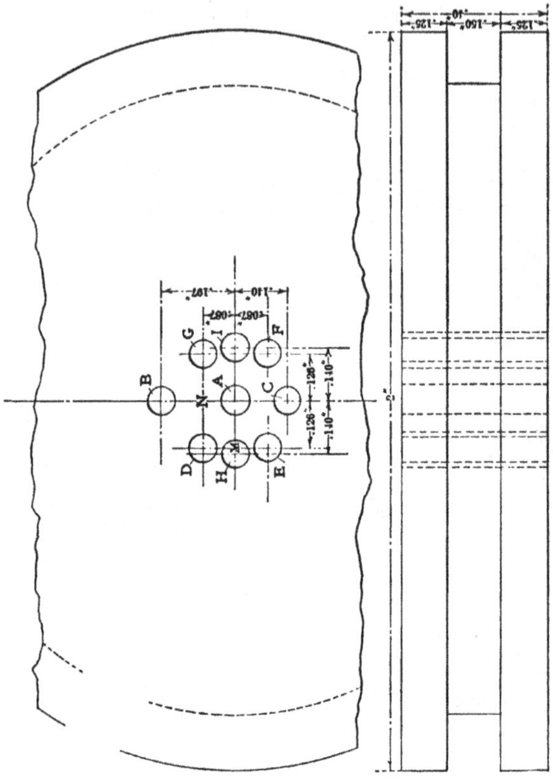

FIG. 80. — CIRCULAR MASTER PLATE

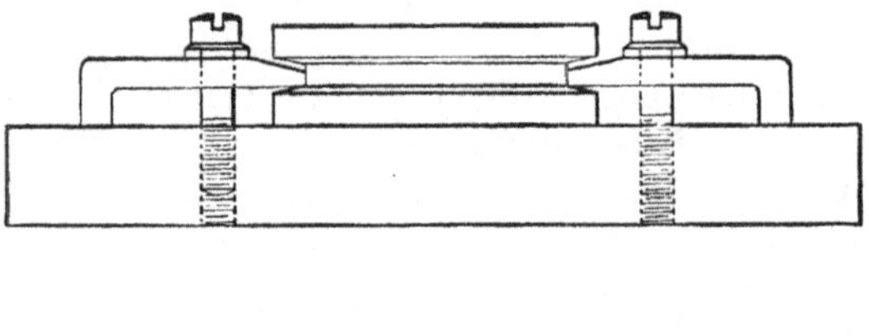

FIG. 79

FIG. 78

RIGHT AND WRONG WAY TO HOLD A MASTER PLATE

THE BUTTON, PLUG AND MICROMETER

The center of hole *B*, Fig. 80, is located approximately 0.197 inch from *A* with a pair of dividers, and is then drilled and tapped for a $\frac{1}{16}$-inch screw. In *A* we place a hardened and ground plug as shown in Fig. 82. With the tapped hole at the approximate center for *B* we clamp a hardened and ground button, then measure for center distance *A B* as in Fig. 82. The center plug in *A*, which is 0.070 inch diameter and the outside end of which is used for measuring over, must be ground perfectly parallel to its axis and with the diameter exact. The button clamped at

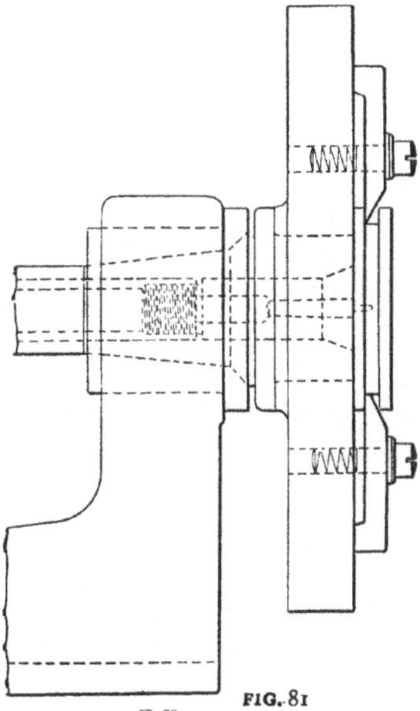

FIG. 81

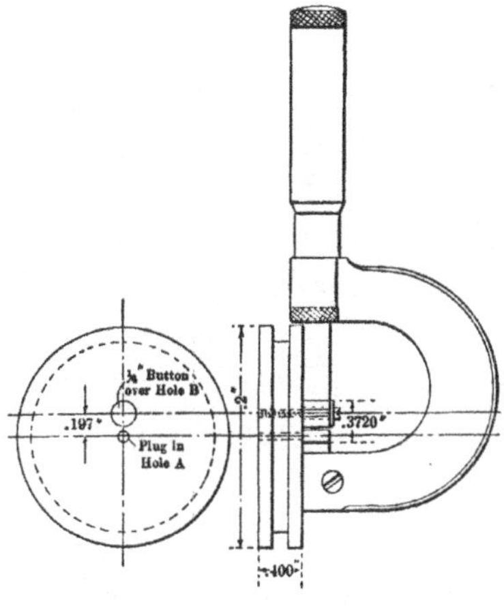

FIG. 82

SETTING THE MASTER PLATE AND LOCATING BUTTON

B must also be hardened and ground parallel to its axis and its ends must be perpendicular to the axis. The button is moved slightly by striking very gently on the side until the micrometer measurement is 0.3720 inch. Half the diameter of the button, 0.125 inch + half the diameter of the end of the plug, which is 0.050 inch = 0.175 inch. 0.3720 inch − 0.1750 = 0.197 inch, which is the distance between *A* and *B*.

We then mount the master plate on the bench-lathe face plate and indicate the button as in Fig. 83. After the button is carefully centered with the indicator, it is removed and hole *B*, Fig. 80, bored and reamed like *A*. The same method is used for locating *C* with the exception that

A, *B* and *C* have to be in perfect alinement. With plugs in *A* and *B* and a button at *C*, all being of equal diameter, we obtain the alinement with a thin, straight edge as in Fig. 84. After indicating the button at *C*, it is removed and hole *C* bored and reamed as were *A* and *B*.

FIG. 83. — INDICATING A BUTTON ON THE MASTER PLATE

FINDING CENTER DISTANCES

Before locating hole *D* we have to find the center distances *B D* and *A D*. The center line through *D G* being perpendicular to the center line through *B A C*, *B N D* forms a right-angle triangle.

$$\sqrt{BN^2 + ND^2} = 0.16726$$

inch or distance *B D*.

$$\sqrt{AM^2 + MD^2} = 0.15312$$

inch or distance *A D*, or radii *A D*, *A E*, *A F* and *A G*. By using a button at *D* and plugs in *A* and *B*, we proceed to locate and bore hole *D* in the same manner as we did with either *A*, *B* or *C*.

By placing a plug in *D* together with the use of a special button as shown in Fig. 85 we will locate hole *E*. Plug *Y* in the center of the special button is made to fit perfectly in hole *A* in the master plate. The distance between centers *O P* in Fig. 85 equals radii *A D*, *A E*, *A F* and *A G*, Fig. 80. Plug *Y*, Fig. 85, is placed in hole *A* and the button set against a plug 0.17176 inch diameter in *D;* the radius of the plug in *D* is 0.08588 inch, which, added to 0.08812 inch, the radius of the special button, = 0.1740 inch or the distance *D E*. The special button is held against the plug in *D* by a small spring and clamped firmly against the surface of the master plate with a screw as shown in Fig. 85.

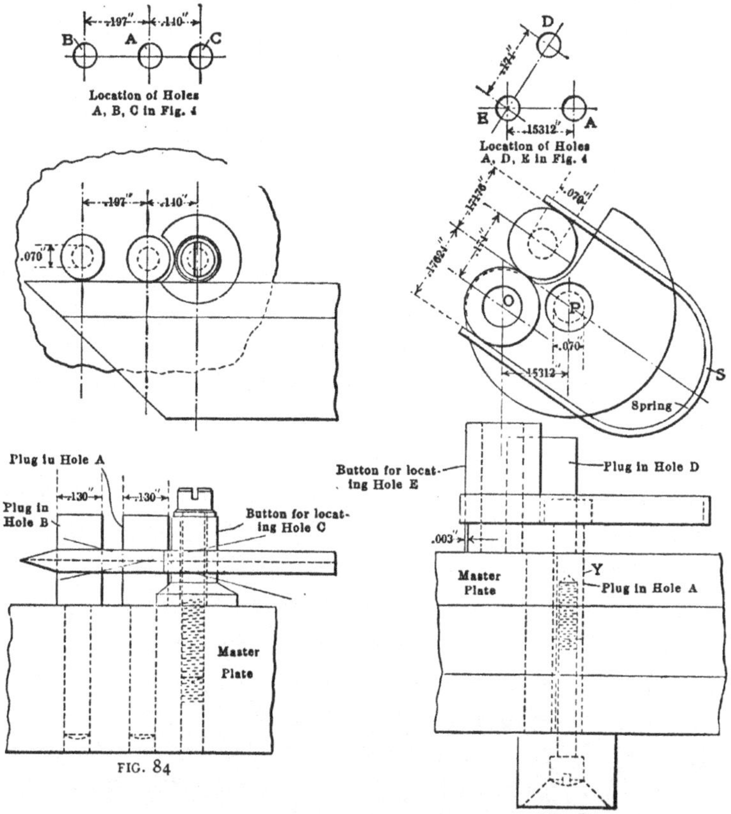

FIG. 84

FIG. 85

LOCATING INDICATOR BUTTON AND FINDING CENTER DISTANCES

INDICATING, BORING AND LAPPING

We then clamp the master plate on the lathe face plate. The special button being longer than the plug in *D* as shown in Fig. 85 enables us to locate *E* by indicating as in Fig. 83. The master plate being securely clamped to the lathe face plate, then the special button, Fig. 85, is removed by putting a screwdriver through the lathe spindle from the back end and removing the screw from plug *Y*. On the head of this screw will be noticed a cupped washer which guides the point of the screwdriver into the screw slot. After taking out the screw, the special button is removed and we proceed to drill and bore hole *E*. Holes *F* and *G* are located in the same manner as *E*. Holes *H* and *I* can be located with a button by measuring from plugs in *B*, *A* and *C*, the same as we located *D* from *A* and *B*, but in this case we have three points to measure from, *A*, *B* and *C*. *H* and *I* can be located with a much greater degree of accuracy by using a special button as in locating *E*, *F* and *G*.

After all the holes are bored and reamed through the plate, they are all lapped just enough to make them of exactly the same diameter. The laps used are generally of brass or copper. The lapping operation is done by holding the master plate in one hand and working the lap with the fingers of the other.

INDEXING ATTACHMENT AND MICROMETER DEPTH GAGES

Fig. 86 illustrates a method used for laying out a master plate where extreme accuracy is not necessary. With the indexing attachment on

FIG. 86. — LOCATING HOLES WITH THE MICROMETER DEPTH GAGE

the bench-lathe head we can get any division of the circle required. The milling attachment with its horizontal and vertical slides enables us to point off any radius required. The movement of the slides in either direction is measured by micrometer depth gages. The gage for measuring the movement of the vertical slide is clamped to an adjustable stop which is held to the knee of the milling attachment by a binding bolt in a T-slot. The depth gage for use on the horizontal slide is clamped to the bed of the lathe by means of a small angle iron. In the end of the spindle next to the face plate is a sharp point or center punch. After setting the point in any position required it is pressed into the surface of the master plate by the movement of the cross horizontal slide. In this way any number of holes can be located approximately correct to the required dimensions.

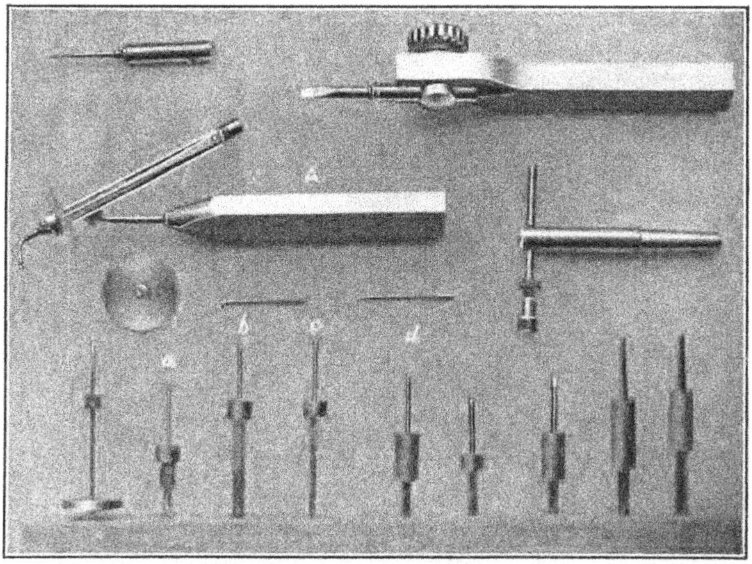

FIG. 87. — TOOLS FOR BORING MASTER PLATES

After the holes in the plate are all marked the prick-punch impressions are indicated with an ordinary indicator; but by using a special arbor with a fine-pointed male center at one end and a female center at the other, and placing the pointed center of the arbor in the prick-punch mark and the female center on the tailstock center, the prick-punch marks can be located with the fine indicator shown at *A*, Fig. 87. In indicating with the special arbor and fine indicator the point of the indicator is placed on the end of the arbor close to the master plate and upon rotating the work the indicator will show the amount the prick-punch marks are out of true. The holes are bored, reamed and lapped in the same way as for the master plate in Fig. 80.

MASTER PLATE RECESS AND BUSHING

Fig. 88 illustrates two points in particular to which it is desired to call attention. The bushing and recess indicated at *B* and *C* show two ways of locating two or more centers when they come within the diameter of a standard hole in the master plate. The recess *C*, which is made to fit the end of a centering plug, in most cases is practicable only for two centers; where three or more centers are necessary within so small a space the bushings should be used. The chief objection to the recess is on account of the necessary changing of the centering plug in the lathe spindle every time the plate is changed from a hole to the recess. If a center plug is once removed from the lathe spindle it cannot in general practice be placed back in the spindle true again, and if the work being done on the plate requires a perfectly true-running center plug it will be necessary to make a new one every time one is removed. With the use of the bushings this is not necessary as the holes in all the bushings can be the same diameter as the standard diameter of master-plate holes.

The master plate in Fig. 88 is equipped with three bushings. One of these, *D*, has the hole concentric, the other two, *B* and *E*, are eccentric. The large hole to receive the bushings is located by using a button as in the case of the master plate in Fig. 80. The bushing blanks are made a plug fit in the hole and keyways are cut in them as shown. A hole is carefully drilled and reamed from the edge of the master plate to the hole that is to be bushed and a pin fitted to this hole with its end passing into the keyway in the bushing blank, thus forming a key for the bushings. The bushing blanks are then set in place one after another and the hole for each bushing is located with a button.

AN ACCURATE GROOVING OPERATION

In Fig. 89 is shown a master plate made especially for putting the grooves in hemispherical piece *F*. This master plate is a T-shaped casting to be clamped to the lathe face plate by screws *G* and *H*. Plate *I* is held at right angles to the lathe face plate and centered and swiveled on the casting by a stud which is held in place by the nut *J*. In the bottom of plate *I* are 10 holes spaced to correspond with the spacing of the grooves required in piece *F*. Plate *I* can be indexed from one hole to another and held firmly in place by index pin *K*.

In mounting the work *F* on plate *I*, it is first fastened to a brass chucking piece, not shown. The chucking piece is held on *I* with screws and a dowel L. Plate *I* is then set in place on the T-iron and left just free enough on the stud to swivel. The center of the centering stud for *I* is in a direct line with the center of hole *M*. This hole fits a center plug in the lathe spindle; the T-iron being clamped to the face plate with

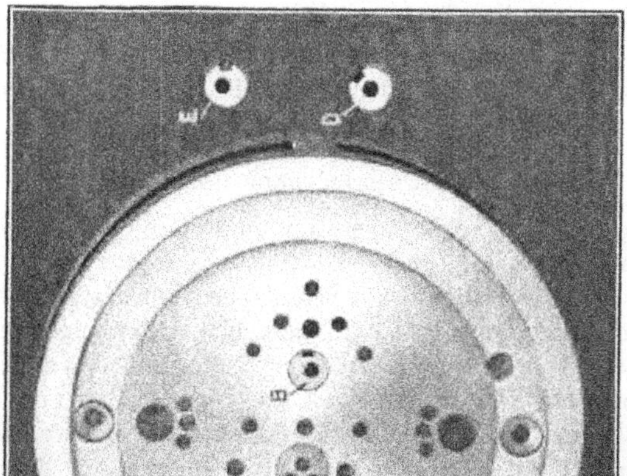

FIG. 88. — A RECESSED AND BUSHED MASTER PLATE

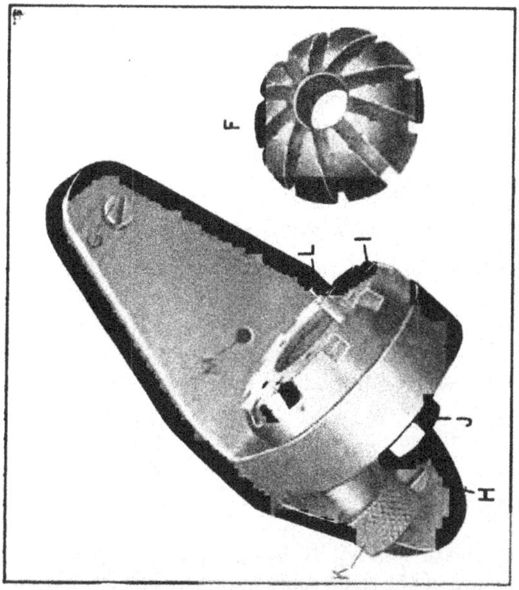

FIG. 89. — A MASTER PLATE FOR A GROOVING OPERATION

screws *G* and *H*. The work is oscillated on the lathe spindle and the grooves milled concentric with the center of hole *M*.

If the piece *F* is to be hardened and ground the above operations are repeated for grinding as for milling. This has proved a very successful way of getting the grooves in piece *F* which have to be very accurately spaced. A great advantage is that the piece can at any time be reproduced without difficulty.

OTHER INDEXING PLATES

Fig. 90 shows another form of master plate which is very useful. Plate *N* has one or more holes as may be necessary. These holes are located to accommodate the work, which is clamped on index plate *O*; the holes in plate *N* are used on the center plug in the lathe spindle. Index plate *O* is centered in plate *N* by a boss on the under side fitting into a recess in *N*. The work being fastened to the index plate, any hole in *N* can be transferred into the work any number of times or the holes spaced any number of degrees apart by the notches cut in the index plate *O*. The index plate is located by latch *P* and held in position by clamps *Q* and *R*.

The master plate in Fig. 91 forms quite an unusual combination. Plates *S* and *T* are of the same combination as Fig. 90, but have an additional movement on main plate *U*. The work is fastened to index plate *S*. This master plate was made for a very intricate piece of work and the illustration simply shows to what extent it is sometimes necessary to complicate such devices in order to accomplish the work in hand.

BORING SMALL, DEEP HOLES

Fig. 92 represents a boring operation on a piece of work mounted on the master plate, shown in Fig. 80. Here the master plate holds the mounting piece in which the work is held. The mounting piece is attached to the master plate with screws. In cases like this, where holes of small diameter have to be carried to considerable depth, the holes are first drilled with a flat or twist drill leaving stock for boring. They are then bored for a short distance just large enough in diameter to receive the end of a hog-nose reamer. The reamer is then fed to the required depth of hole. Several reamers are seen at *a*, *b*, *c*, *d*, Fig. 87. This operation is repeated four times, removing with each reamer from the hole about 0.003 inch until the hole is the correct diameter.

This method works well in a deep hole of small diameter and is much quicker than using a single-pointed boring tool in the slide rest for the full depth of the hole, and with care the hole can be made as true as with the single-pointed boring tool. The collars, or rings, on the reamers in Fig. 87 are used for stops where holes are to be of a certain depth. These

FIG. 90. — AN INDEXING PLATE

FIG. 91. — ANOTHER INDEXING PLATE

stops are simply steel or brass collars soldered to the reamer, and with the end faced off until the required distance from the point of the reamer to the stop collar is obtained.

FIG. 92. — BORING SMALL DEEP HOLES

LOCATING AND MILLING A SLOT

Fig. 93 shows the method of milling a slot in a piece *D*. This slot, which has to be perfectly radial to the hole *E*, will be seen at *F* behind the cutter, and is more clearly shown in Fig. 94 which illustrates the method of preparing the master plate *G* for holding the work D in correct position.

The two plugs *H* and *I*, of equal diameter and held in the brass mounting piece *J*, Fig. 94, are placed in holes transferred from master plate *G*. Plug *H* is the correct center for hole *E* and plug *I* is in the correct position to bring the center of slot *F* in the proper relation to other holes in the master plate. *I* is also the center for the outside of the piece of work. The master plate *G* is placed on a centering plug in the lathe spindle, at the center of plug *I*. The spindle of the lathe is then indexed until the indicator shows plugs *H* and *I* to be parallel to the movement of the horizontal slide. Mounting piece *J* contains a recess concentric with plug *I*, which was used for mounting the work *D* at the time the hole *E* was bored. Plugs *H* and *I* are removed and the work is again clamped in the

FIG. 93. — A SLOT MILLING OPERATION

FIG. 94. — PREPARING THE MASTER PLATE FOR THE OPERATION IN FIG. 93

recess which is concentric with *I*. Hole *E* is held concentric with the center of *H* by a special plug fitting *E* and the hole from which plug *H* is removed.

After clamping the work and placing a cutter in the spindle of the milling attachment we can proceed to mill the slot *F*, as in Fig. 93. The slot is first rough-milled, and then to bring it central to the center line through plugs *H* and *I* we take a light cut on one side, then index through 180 degrees and take a cut on the other side without changing the setting of the cutter, and repeat this operation until the slot is the required width.

MILLING AND GRINDING OPERATIONS

Fig. 95 illustrates another interesting milling operation. The irregular shaped die or mold fastened to the face plate is worked out from holes in the master plate to which it is clamped.

FIG. 95. — MILLING OUT A DIE

Fig. 96 shows an operation of grinding four holes in a piece of work. The holes were transferred into the work before hardening, on the same master plate, and after hardening it was remounted for grinding them to again match the master plate. The master plate is located on the face plate for each hole in the usual way by a centering plug in the lathe spindle. The lathe head and the slide rest holding the traverse spindle grinder are set high above the lathe bed on raising blocks so that the weights used for counterbalancing the work will clear the lathe bed. In the end of the traverse spindle is mounted a steel lap charged with diamond powder for grinding the holes.

FIG. 96. — A GRINDING OPERATION

DIE-MADE CASTINGS

Fig. 97 should be of interest as illustrating some of the castings produced by the Veeder Manufacturing Company's process for making small parts that would be difficult and very expensive to machine. This class of work is all made in dies produced by the aid of master plates as represented in this chapter. The casting at *A*, for example, is made in the die shown on the face plate in Fig. 95.

FIG. 97. — VEEDER DIE CASTINGS

CHAPTER VIII

Master Plates Used in Making Watch Tools

In getting out the necessary tools for the manufacture of a watch movement after the master watch or model has been constructed with every dimension as accurate as the skilled mechanic with a perfect equipment can make it, a set of master plates is indispensable. A group of work plates for one of the movements manufactured by the E. Howard Watch Company, Waltham, Mass., is shown in the foreground of Fig. 98, while some fixtures closely identified with the plates are represented at the rear and to the right in the engraving.

MASTER AND REFERENCE PLATES

In these master plates are holes corresponding to the wheel-train positions in the watch, and giving the locations of various screws and pin holes, the centers for recesses, curves, and fillets to be swept out. No one plate contains the entire series of holes, as some of them are very close together or actually overlap. There are, however, a number of holes that are repeated in different plates; for example, three near the periphery which locate the three main post or pillar holes in the watch plate and which may be considered the matching holes in the different master plates. The disk at *A* is a master reference plate in which are the three holes just referred to, and other holes whose positions must be correct; these including, of course, the ones that determine the train positions, correct center distances between, which are of prime importance. The plate *A* is used solely for reference purposes and hence forms a permanent gage by which the accuracy of the master plates and tools made from them may at any time be tested. The reference plate, and the regular master plates when not in use, are kept in a safe, and naturally when used are handled with care and judgment in order that their accuracy may be preserved.

MASTER-PLATE FIXTURE

In transferring holes from a given master plate to another plate the holders of the types shown at *B* and *C* are of service. A few dimensions are given in the sketch in Fig. 99. After a blank disk like *D*, Fig. 98, has been drilled, as shown in jig *E* for a single locating pin, it is dropped into the chamber in one side of the ring-shaped fixture *B* in which it is a good

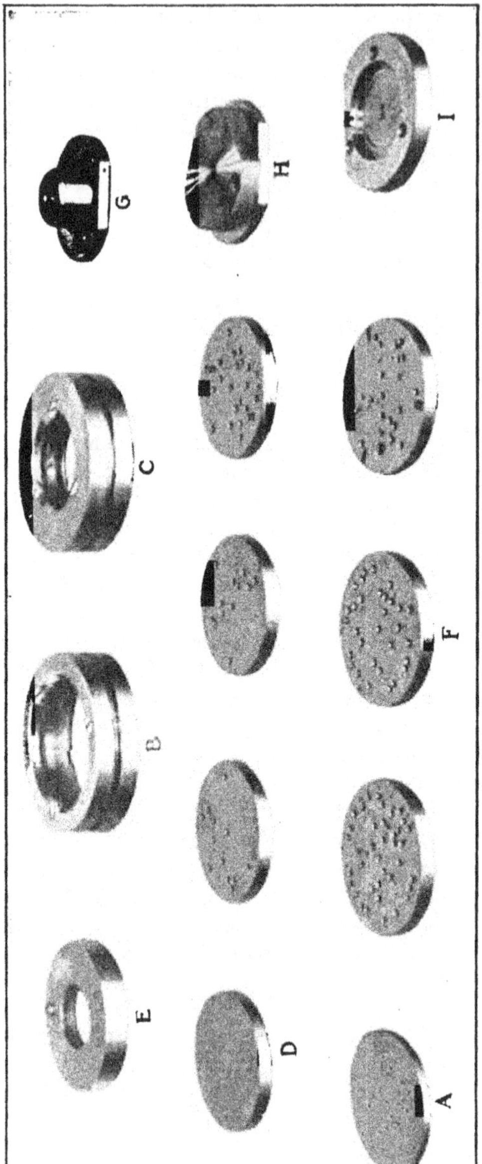

FIG. 98. — MASTER PLATES FOR A HOWARD WATCH

fit and where it is located by the dowel pin and fastened by three fillister-head screws. The master plate is placed in the opposite side of *B*, secured by three screws, and the whole is then ready to be mounted on the bench-lathe face plate, or on a face-plate quill in a regular bench-lathe quill rest, where one hole after another is bored through the blank plate, the work being located for each hole by slipping the master plate over a centering plug projecting from the spindle, which plug is made true and a perfect fit for the master-plate holes. Where the plate being bored out is to be

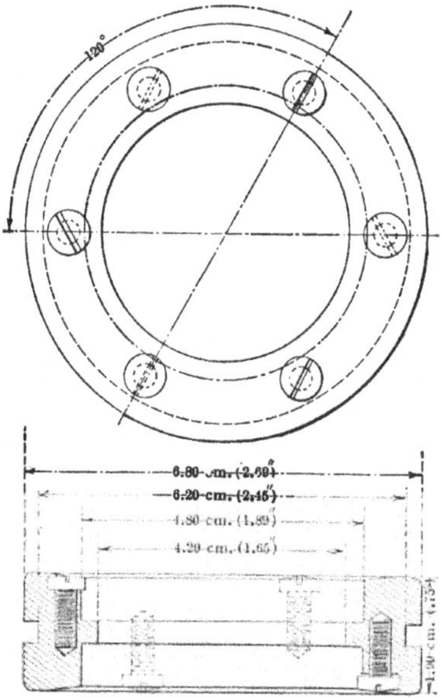

FIG. 99. — MASTER-PLATE HOLDER

hardened the holes can be bored undersize and then, after hardening, each hole can be ground out to size with a lap, using the same method of locating as when boring the holes. The groove formed in the periphery of the fixture allows it to be clamped readily to the face plate by means of light straps. The master plates, six of which are shown in the central group *F*, and the reference plate *A*, are made to the same diameter, which is practically 2 inches.

The method of locating the work on the face plate with the master plate on the centering plug in the spindle needs no special illustration here, as it has been shown in Chapters VI and VII. The bench-lathe quill rest is also familiar to readers; this being a supplementary head which is placed in front of the regular headstock and adapted to receive the quill and its face plate, whose spindle running in the quill is driven by a connecting member from the regular head spindle. This arrangement is described in Chapter VI, and the use of the central plug in the spindle for locating the master plate is illustrated in the same chapter.

FACE-PLATE DEVICES

The appliances at *G*, *H* and *I* are used for mounting brass holders in which pieces of various shapes may be secured while boring out or working to correct outline on the face plate. Devices of this character may be used if desired in a holding fixture of the type shown at *B*, with a master plate in the underside for locating the work correctly for the different centers to which the piece is to be bored or otherwise machined; or where the work is to be done without a master plate, as the disks of the form shown at *G H I* are flatted on the edge, they may be conveniently placed on a parallel strip on the face plate and adjusted by size blocks or wires to give the positions required for boring the holes in the work which they carry, or for working out arcs, recesses, etc. A second strip may be placed at right angles to the one upon which the disk rests, and after a hole is bored in the work, wires of the required size may be introduced between the parallels and the work-holding fixture and the latter reclamped for the boring of the next hole. The brass blocks fastened to the steel holders can be of any convenient form and size and bored out to receive punches, or any small pieces requiring accurate machining. They may also be used as chucks for mounting thin, flat work by means of soft solder or shellac.

EQUIPMENT FOR MASTER PLATE AND OTHER WORK

In Fig. 100 is shown some of the equipment utilized in connection with the making of master plate and other precision tools. The block carrying the short pieces of wire is drilled (and stamped) for a large number of sizes of small wire, the variation throughout the range being by very minute increments. With these wires used as setting gages in the manner mentioned in the preceding paragraph, the work may be accurately adjusted to the different positions required on the face plate as one hole after another is bored.

The application of button and disks in the locating of holes in particular work is well understood by toolmakers and by others who have followed the preceding chapters. Both disks and buttons are used extensively in conjunction with the indicator when performing the operations now under consideration, and a number of these indispensable appliances are shown in Fig. 100. Many of the disks, as will be noticed, are notched or flatted at the periphery to permit of the securing of the required center distances.

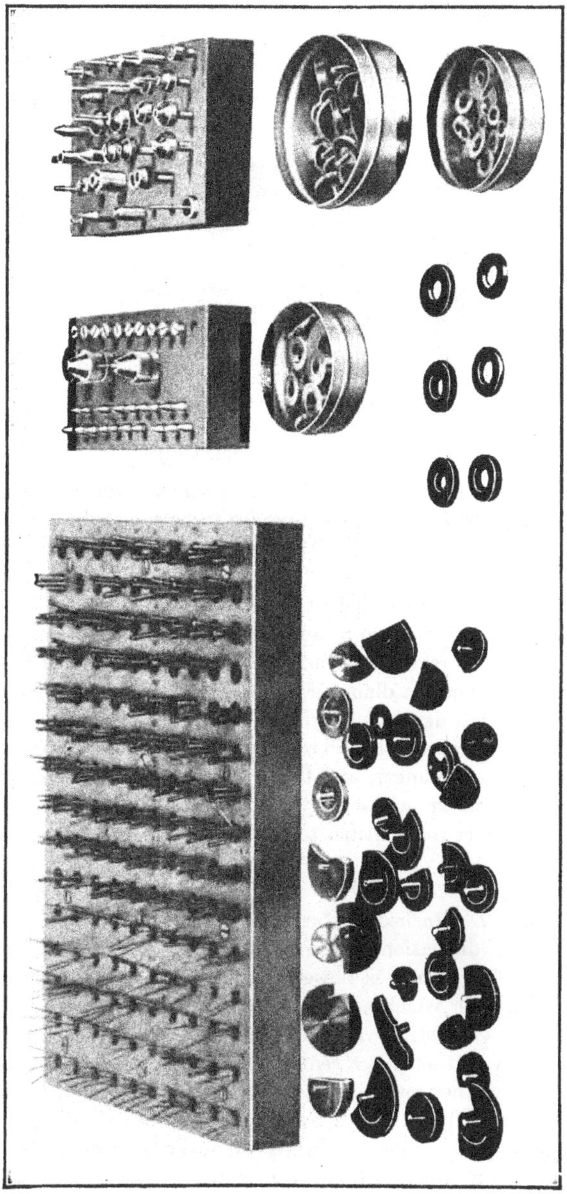

FIG. 100. — HANDY EQUIPMENT FOR THE TOOLMAKER

The sets of small chucks, collets and holders in the blocks at the upper right-hand corner are found useful for various operations arising in connection with watch-lathe processes. Details of a chuck and collet are given in Fig. 101, with a few approximate dimensions in inchcs.

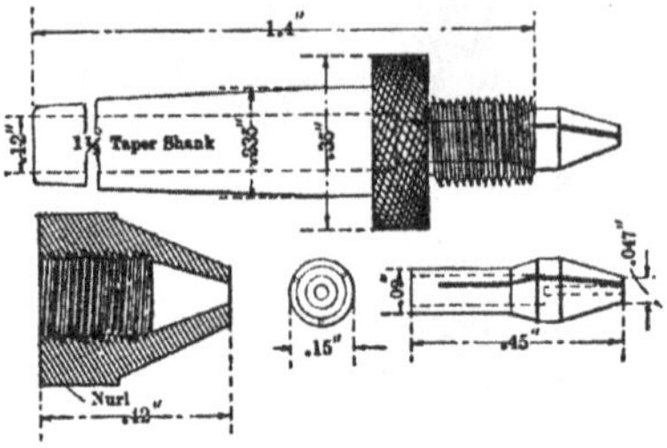

FIG. 101 —SMALL DRILL CHUCK

HANDY TOOLS AND APPLIANCES

In originating master plates or handling similar problems where laying out, pricking centers and indicating are to be resorted to some of the tools shown in Fig. 102 are convenient; different appliances here illustrated are in daily use among the Howard toolmakers.

The instrument at *A* is a neat little device for scribing circles on metal and somewhat resembles in appearance and operation a small bow pen. The scribing blade is revolved about the stem by the finger and will produce a circle of as small a diameter as required. The tool at *B* is also a handy and ingenious device for scribing circles and arcs concentric with a hole. The pump center when in operation is held in its seat by pressure of a spring behind a plunger, and the scribing point has a quick though sensitive adjustment the whole length of the beam, by a small screw with rapid lead. Another tool constructed on somewhat similar lines is shown at *C*.

ACCURATE INDICATORS

The indicator *D* attached to the surface gage and the one shown at *E* are made to the dimensions given on the indicator drawing shown in Fig. 76, Chapter VI. There is, however, a difference in the spring arrangement under the lever which actuates the pointer. The spring is double, that is, there are two spring wires extending under the lever, and the ends of these normally rest against a centering pin fixed in the indicator body; at the same time they press against the pin which connects indicator pointer and lever. Thus they are always under initial tension, and no matter which way the pointer is swung when the indicator is in use it is brought back to central position when contact with the work is interrupted.

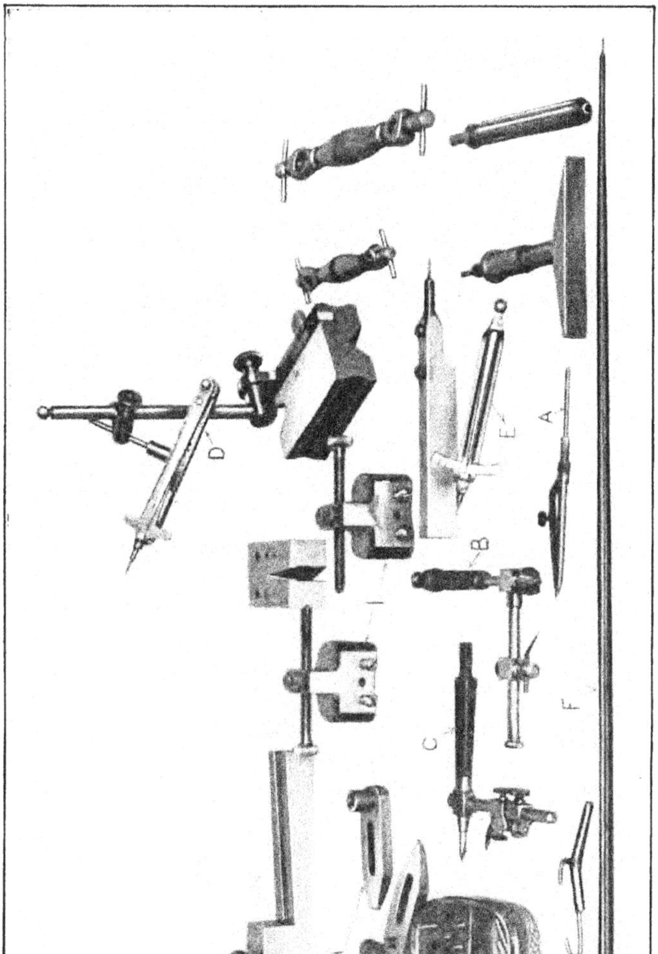

FIG. 102. — APPLIANCES FOR FINE TOOL WORK

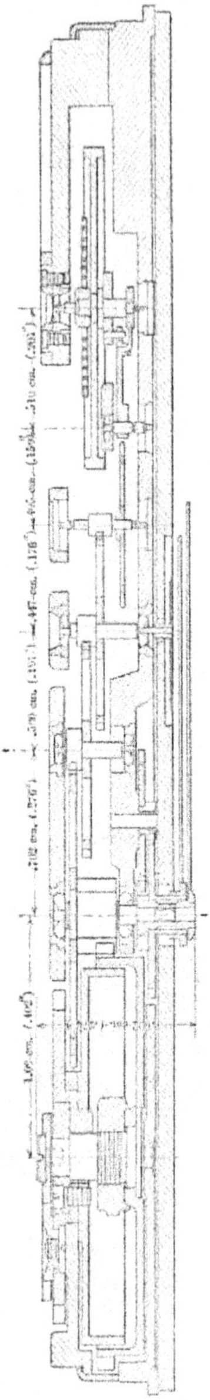

FIG. 103.—SECTION AND TRAIN DEVELOPMENT OF HOWARD 12 SIZE, EXTRA THIN MODEL WATCH (SCALE 3 TO 1)

Another type of indicator is shown at *F* with the attachments for mounting it immediately in the rear. A variety of indicator points will be noticed in box *G*.

CLAMPING AND OTHER DEVICES

The small straps at *H* for securing face-plate work are used in conjunction with the tapped plugs shown. These are made of different lengths to suit various thicknesses of work, and are attached by screws to the rear ends of the small straps, making it unnecessary to use blocks or other loose pieces under them.

The attachments *I* are intended to be clamped to the flange of a bench-lathe pulley, one at the front and one at the rear, with the adjusting screws resting against the side of the head. They may then be used for locating the spindle or turning it backward or forward through a certain distance determined by the character of the operation to be performed. The adjustable hight blocks at the rear, the V-block for use in removing or setting small arbors, and pins, and the various tool holders, micrometer depth gage, etc., are clearly shown and need no explanation.

A THIN WATCH MOVEMENT

As a specific illustration of a piece of work, certain tools for which are made by the aid of master plates of the type described in this chapter, the

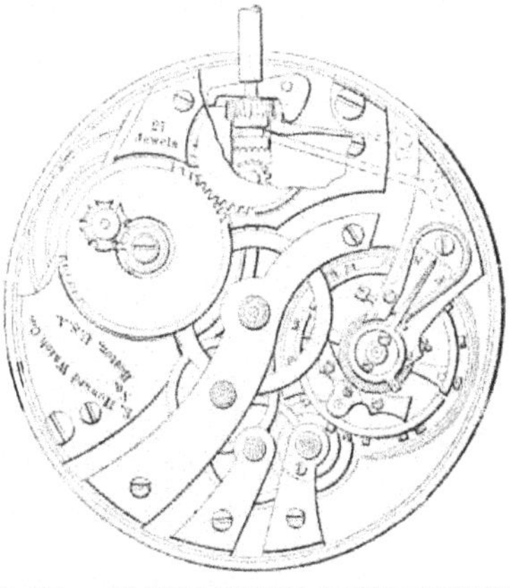

FIG. 104. — PLAN OF HOWARD WATCH MOVEMENT (SLIGHTLY ENLARGED)

watch movement in Fig. 103 should be of interest. This drawing shows on a scale of about 3 to 1 the construction of the new "12 size" Howard extra thin model watch, which, so it is stated, is the thinnest gentlemen's watch made. A few dimensions are included in the engraving to show, among other things, actual center distances from wheel to wheel. The arrangement of the jewels is also shown in this drawing.

It should be noted that the sectional view does not represent a section straight across the diameter of the watch movement, but is instead what may be termed a zig-zag section from center to center along the train, opened out into a common plane. A plan view to smaller scale is given in Fig. 104.

CHAPTER IX

Trigonometry in the Tool Room[1]

Very likely the title of this chapter will frighten many from reading it, though there is no reason why it should. Triognometry is looked upon much as algebra is, but more so, and a glance into a text-book on the subject will justify the impression that it runs into somewhat fearsome looking equations. This does not, however, alter the fact that many useful applications of trigonometry are very simple, involving, as they do, nothing more than taking certain quantities from tables and multiplying them by dimensions laid down on drawings in order to obtain others which are more convenient for the toolmaker. The whole thing is so simple that if the word trigonometry could be left out of the title, or, if the title could be placed at the bottom of this chapter, many an unsuspecting person would read it without supposing that he was doing anything more than learning the use of certain tables to obtain certain results for which he had before been obliged to go to the drawing room and consult some one with a wrinkled brow and a reputation as a "mathematician" — a reputation based on nothing more, in instances like those here cited, than the fact that, his methods being unknown, they are assumed to be profound and complex, when in fact they involve nothing more than simple multiplication, and not much of that. The present article does not contain a single equation nor a Greek letter, and this not because they have been forcibly suppressed, but because they are not needed.

CHANGING AWKWARD TO CONVENIENT DIMENSIONS

Draftsmen have a fondness for laying out various points in a jig, for example, by angles expressed in degrees, and when a drawing of this kind involving really accurate work comes to the shop, some one must change the angle readings to linear measurements from base lines, and the toolmaker who can do this can very readily get the reputation of being deeply learned.

Suppose a jig drawing comes to the shop with the locations of two holes dimensioned as in Fig. 105. With the micrometer dial of his milling machine feed screw the toolmaker may determine the base distance with a good degree of accuracy, but he has no means of laying down the angle

[1] F. A. Halsey in the *American Machinist.*

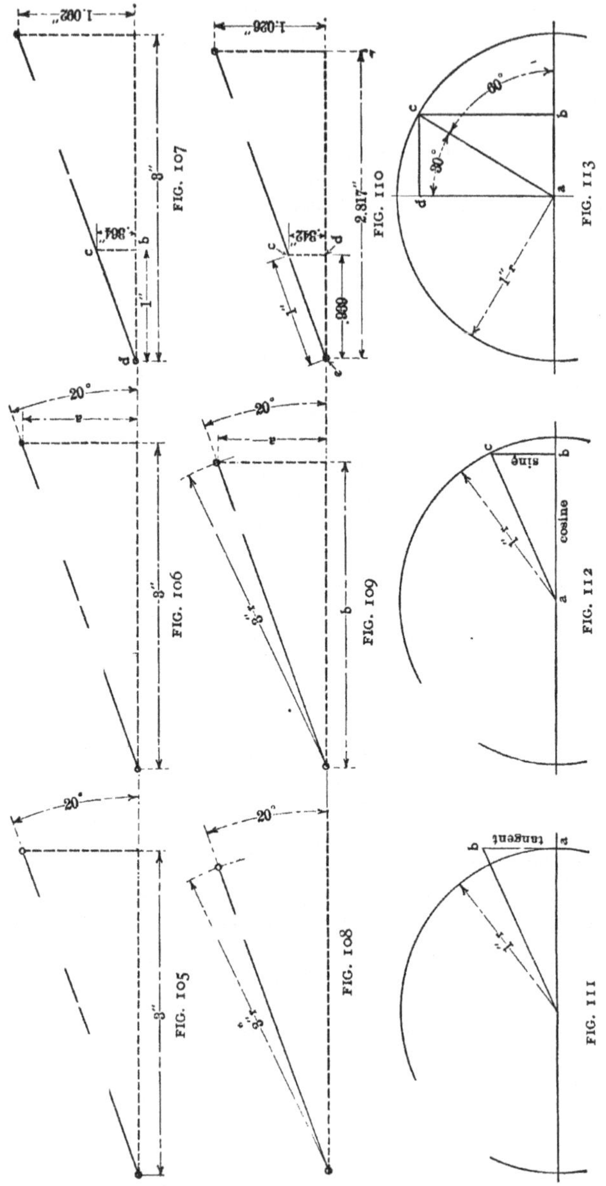

FIG. 105 FIG. 106 FIG. 107

FIG. 108 FIG. 109 FIG. 110

FIG. 111 FIG. 112 FIG. 113

CHANGING ANGULAR TO LINEAR MEASUREMENTS

with any such precision, whereas if the distance *a*, Fig. 106, could be determined it could be substituted for the angle reading, and, by using the vertical feed screw, be laid out just as readily as the base distance.

To do this all that is necessary is to look in a suitable table for the length *b c*, Fig. 107, of a triangle having a base *d b* of one inch and a base angle of 20 degrees. The table gives this length as 0.364 inch, as noted on the drawing, and multiplying it by 3 we find the value of *a*, Fig. 106, to be 1.092 inches, and we re-figure the drawing, as in Fig. 107, and this is all there is of it. The table gives a complete list of the lengths of *b c* for all angles between 0 and 90 degrees and for a length *d b* of one inch, or one foot if we are dealing with feet. Had the angle been 30 degrees we would have found the length of *b c* for that angle from the table, and so for any other angle.

Drawings often come to the shop dimensioned as in Fig. 108. With his button process and micrometer the toolmaker may determine the radial distance between the holes with a high degree of accuracy, but again he is unable to lay down the angle with precision and it is necessary to obtain the distance *a* of Fig. 109, which may be readily measured. The base distance *b*, Fig. 109, is also more convenient for the toolmaker than the radius, and while we are about it, we may as well find *b* and then discard the radius.

To do this all that is necessary is to look in other columns of the same table for the lengths of *c d* and *e d*, Fig. 110, of a triangle having a slope of one inch and a base angle of 20 degrees. The table gives these lengths 0.342 and 0.939 inch respectively, as noted, and multiplying them by 3 we find the value of *a*, Fig. 109, to be 1.026 and of *b* to be 2.817 inches, and again we re-figure the drawing as in Fig. 110.

The only effect of these changes is to give figures which are more convenient for the toolmaker, the layout of Fig. 107 being exactly equivalent to that of Fig. 105, while that of Fig. 110 is exactly equivalent to that of Fig. 108.

THE NAMES USED

The columns of the tables giving the sides of the small or unit triangle are headed *tangent*, *sine* and *cosine*, and all that remains is for the reader to learn the meanings of these words in order that he may use the proper column in each case.

In Fig. 111 the base of the unit triangle or the radius of the circle is one inch, under which circumstance the length of the line *a b* in fractions of an inch is the *tangent* of the angle. Comparing Figs. 107 and 111 it is obvious that the value 0.364 inch of Fig. 107 is the tangent of 20 degrees, and if the reader will turn to a table of sines, cosines and tangents he will find this value in the tangent column opposite 20 degrees. For this purpose see the skeleton table on page 100.

Turning next to Fig. 112, the unit triangle has its slope or the radius of the circle equal to one inch, under which circumstance the length of the line *b c* in fractions of an inch is the *sine* of the angle, and similarly, the length of *a b* is the *cosine* of the angle. Comparing Figs. 110 and 112 it is obvious that the value 0.342 inch, Fig. 110, is the sine of 20 degrees, and if the reader will turn to a table of sines, cosines and tangents, he will find this value in the sine column opposite 20 degrees. Similarly the value 0.939 inch, Fig. 110, is the cosine of 20 degrees and, as before, the reader may find this value in the cosine column opposite this angle. Again use the skeleton table for this purpose.

It will be observed that the sine gives the length of the rise as compared with the length of the *slope*, while the tangent gives it as compared with the length of the *base*. If the original drawing gives the length of the slope, the column of sines is to be consulted, while if the length of the base is given, the column of tangents is to be used.

Tables of sines, cosines and tangents may be found in any engineer's pocket-book. In this connection it may be well to warn the reader that the tables to be used are those of *natural* sines and tangents, not those of logarithmic sines and tangents, which are very different things, of no value in this connection.

HOW TO USE THE TABLES

One feature of the tables by which their length is halved remains to be explained. Referring to Fig. 113, the *cosine a b* of 60 degrees will be seen, by turning the page 90 degrees to the left, to be equal to the *sine c d* of 30 degrees, and this is obviously true for all complementary angles. We might therefore cut the table in half at 45 degrees and find the cosine of 60 degrees by looking for the sine of 30 degrees and so for any angle above 45 degrees. This would, however, involve subtraction, and to obviate this, two sets of figures for angles are given in the tables.

SKELETON TABLE OF SINES AND TANGENTS

Degrees	Sine	Tangent	Cotangent	Cosine	Degrees
0	0.0000	0.0000	Infinite	1.0000	90
5	0.0872	0.0875	11.430	0.9962	85
10	0.1736	0.1763	5.6713	0.9848	80
15	0.2588	0.2679	3.7320	0.9659	75
20	0.3420	0.3640	2.7475	0.9397	70
25	0.4226	0.4663	2.1445	0.9063	65
30	0.5000	0.5773	1.7320	0.8660	60
35	0.5736	0.7002	1.4281	0.8191	55
40	0.6428	0.8391	1.1918	0.7660	50
45	0.7071	1.0000	1.0000	0.7071	45
	Cosine	Cotangent	Tangent	Sine	

This is shown in the skeleton table herewith, using which to find the sine or cosine of an angle less than 45 degrees we read the angle from the left-hand column and the sine and cosine from the captions at the heads of the columns, while for angles greater than 45 degrees we read the angle from the right-hand column and the sine and cosine from the captions at the foot of the columns. Thus we have for the sine of 25 degrees, 0.4226, and for its cosine, 0.9063, while for the sine of 65 degrees we have 0.9063, and for its cosine, 0.4226.

Precisely so again the tangent of an angle is called the *cotangent* of its complement, and the tangent of an angle is found in the same way as the sine — from the left-hand figures and the caption at the top, if the angle is less than 45 degrees, and from the right-hand figures and the caption at the bottom of the column, if the angle be greater than 45 degrees. Thus we have for the tangent of 25 degrees, 0.4663, and for the tangent of 65 degrees, 2.1445.

CHAPTER X

Tool for Laying Out Angles [1]

A CONVENIENT tool for laying out angles is shown about half size in Fig. 114. In using this gage set a vernier caliper or large micrometer to

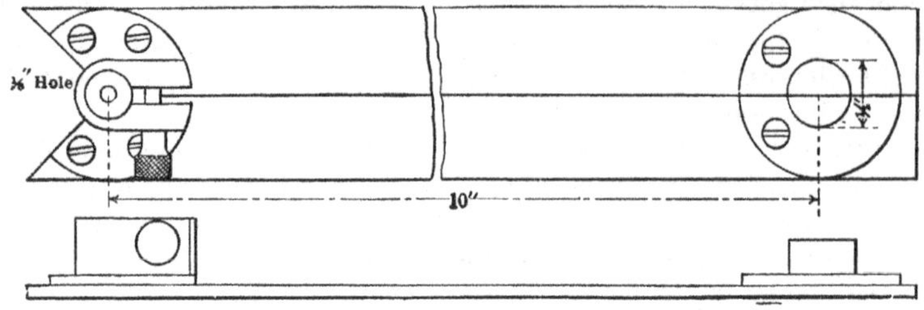

FIG. 114. — TOOL FOR LAYING OUT ANGLES

twice the sine of half the angle desired multiplied by ten, adding one-half inch, and open the gage till it fits the vernier; this will give the angle within the limits of the measuring tool and the radius of the gage. The eighth-inch hole in the center is for a setting plug when it is desirable to lay out an angle from a given center.

TABLE FOR SETTING TOOL FOR LAYING OUT ANGLES

Gage Settings for Even Degrees

Angle Degrees	Measurement Over Disks	Angle Degrees	Measurement Over Disks	Angle Degrees	Measurement Over Disks	Angle Degrees	Measurement Over Disks
1	0.6746	12	2.5906	23	4.4874	34	6.3474
2	0.8490	13	2.7640	24	4.6582	35	6.5142
3	1.0236	14	2.9374	25	4.8288	36	6.6804
4	1.1980	15	3.1106	26	4.9980	37	6.8460
5	1.3724	16	3.2834	27	5.1690	38	7.0114
6	1.5468	17	3.4562	28	5.3384	39	7.1762
7	1.7210	18	3.6286	29	5.5176	40	7.3404
8	1.8952	19	3.8010	30	5.6764	41	7.5042
9	2.0692	20	3.9730	31	5.8448	42	7.6674
10	2.2432	21	4.1448	32	6.0128	43	7.8300
11	2.4170	22	4.3162	33	6.1804	44	7.9922
						45	8.1536

[1] Contributed to the *American Machinist* by Walter Dillaway.

GAGE SETTINGS FOR HOLES IN A CIRCLE

No. of Holes in Circle	Measurement Over Disks	No. of Holes in Circle	Measurement Over Disks	No. of Holes in Circle	Measurement Over Disks	No. of Holes in Circle	Measurement Over Disks
3	17.8206	8	8.1536	13	5.2864	18	3.9730
4	14.6422	9	7.3404	14	4.9504	19	3.7918
5	12.2558	10	6.6802	15	4.6582	20	3.6286
6	10.5000	11	6.1346	16	4.4018	21	3.4808
7	9.1776	12	5.6762	17	4.1750	22	3.3462

The table gives the measurements over the half disks required for setting the arms of the gage to give any angle from 1 to 45 degrees, in even degrees, and also the setting for any number of holes in a circle from 3 to 22; the latter feature, of course, is of service more particularly for work which does not require the accuracy obtainable by means of a dividing head.

CHAPTER XI

Measuring Dovetail Slides, Gibs and V's[1]

The measuring of dovetails is an interesting and important subject, and for this reason a general treatment covering wide applications of this principle of measuring may be welcomed by many machinists and toolmakers.

Certain formulas have been published for finding the maximum diameter of wire that may be used to measure the dovetail, by the method illustrated in Fig. 115. This means that the point of contact with the wire of maximum diameter is on the very upper edge of the dovetail. This condition is not desirable. The line of contact should be kept a little lower, as shown in Fig. 115. Many dovetails have clearance on top and a couple of thousandths less in hight would not make any difference in the fitting of the slide, but would throw out the measurement of the dovetail. The size of wire obtained by the formulas referred to will be an odd one and will have to be made specially. If not made special, the next smaller standard size has to be taken. For this reason the figuring of the size of wire seems to be wasted time. It can be very readily seen from the drawing or with a wire against the working piece, if the wire is the required size.

GAGING FEMALE DOVETAILS

Sometimes the female dovetail is gaged by measuring between the wires, as at *A*, Fig. 116. The more common practice is to measure the outside of the wires, as at *B*. The wires have to project a little at the end of the slide and a micrometer reading can be taken easily. In most cases the male part will be planed first and the female part fitted to it, or if a large number have to be planed, gages will be made on the same principle.

To simplify the matter of measuring dovetails several tables for practical limits have been worked out and are given here. On the smaller sizes of dovetails the diameter of the wire increases by $\frac{1}{16}$ and $\frac{1}{8}$ inch, on the larger ones by $\frac{1}{4}$ and $\frac{1}{2}$ inch.

ORDINARY FORMS OF DOVETAILS

Considering dovetails of the shape shown in Figs. 117 and 118, on the drawing the draftsman will give either dimension *A* or *B* for the width of

[1] Contributed to the *American Machinist* by Frank A. Scheu.

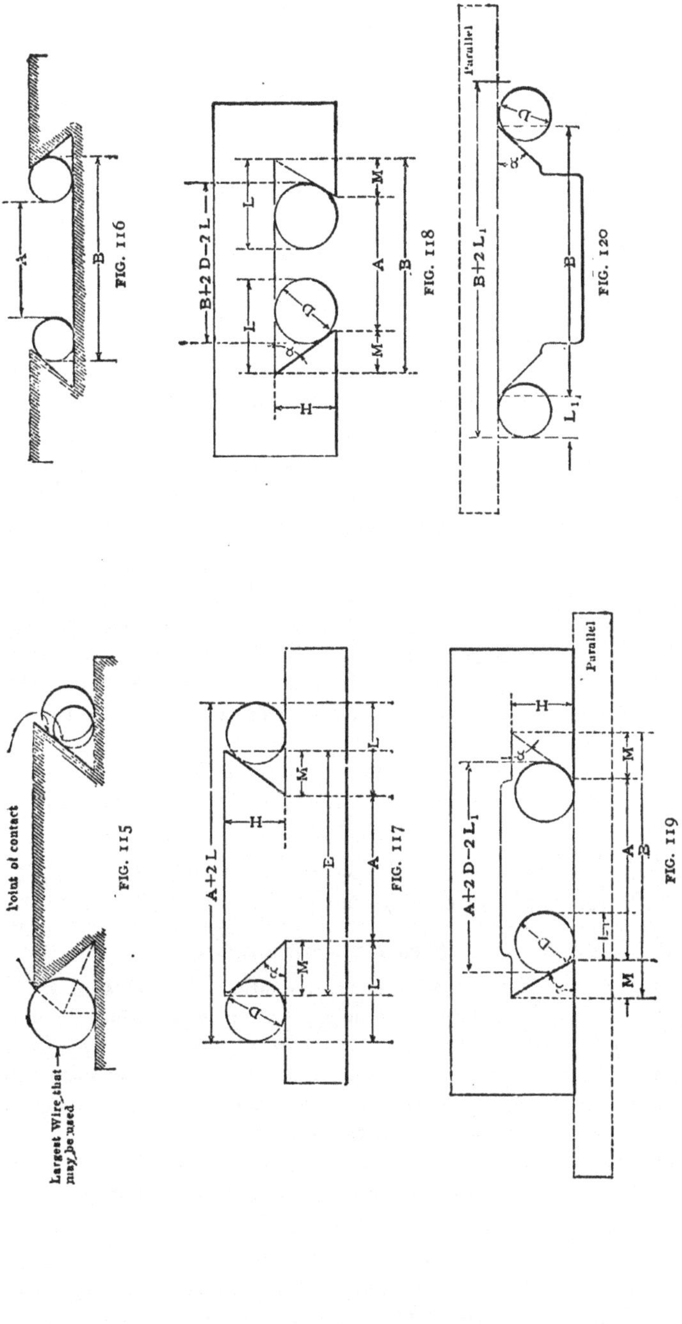

FIG. 115 FIG. 116 FIG. 117 FIG. 118 FIG. 119 FIG. 120

MEASURING DOVETAILS

the dovetail, sometimes both. To find either dimension A or B if the other is given, figure out M according to the formula:

$$M = \frac{H}{tan.\ a}.$$

If angle a = 45° distance $M = H$.

If angle a = 50° distance $M = \frac{H}{1.19175}$.

If angle a = 55° distance $M = \frac{H}{1.42815}$.

If angle a = 60° distance $M = \frac{H}{1.73205}$.

To find L the formula $L = \frac{\frac{D}{2}}{tan.\ \frac{a}{2}} + \frac{D}{2}$ has to be used.

Table 1 has been figured this way and applies to the dovetails shown in Figs. 117 and 118.

THE USE OF PARALLELS

On several dovetails the top of the female slide is either rough planed or not machined at all. In this case the measurement has to be taken from the bottom part. A parallel has to be clamped across the bottom and the reading taken as illustrated in Fig. 119. This is the more correct way of measuring the female dovetail, if clearance is given on top; only it is a little inconvenient and cumbersome. To find L_1 the formula

$$L_1 = \frac{\frac{D}{2}}{tan.\left(90° - \frac{a}{2}\right)} + \frac{D}{2}$$

has been used and Table 2 developed this way. Table 2 also will be of use for dovetails as illustrated in Fig. 120.

APPLICATION TO DOVETAIL GIBS

Figs. 121 and 122 illustrate the measuring of a dovetail gib. In Fig. 121 the top is the important surface and the measurement should be taken in relation to that surface.

Dimension L is required and given under Table 1. If for some special reason it is preferred to work from the bottom, as shown in Fig. 122, Table 2 will give the required dimension L_1.

Fig. 123 illustrates a gage for the gib shown in Fig. 121. The required measurement in this case is K, which can be easily obtained. Use

For Dimension L											
	D = 1/4	D = 5/16	D = 3/8	D = 7/16	D = 1/2	D = 5/8	D = 3/4	D = 7/8	D = 1	D = 1 1/4	D = 1 1/2
α = 45°	.4268	.5335	.6402	.7469	.8536	1.0670	1.2804	1.4938	1.7071	2.1339	2.5607
α = 50°	.3931	.4913	.5896	.6879	.7861	.9827	1.1792	1.3757	1.5723	1.9653	2.3584
α = 55°	.3651	.4564	.5477	.6390	.7303	.9128	1.0954	1.2779	1.4605	1.8256	2.1907
α = 60°	.3415	.4269	.5123	.5976	.6830	.8538	1.0245	1.1953	1.3660	1.7075	2.0490

Table 1. — Values of L (Figs. 117, 118, 121, 124 and 126)

For Dimension L_1											
	D = 1/4	D = 5/16	D = 3/8	D = 7/16	D = 1/2	D = 5/8	D = 3/4	D = 7/8	D = 1	D = 1 1/4	D = 1 1/2
α = 45°	.1768	.2210	.2652	.3094	.3536	.4419	.5303	.6187	.7071	.8839	1.0607
α = 50°	.1833	.2291	.2749	.3208	.3666	.4582	.5499	.6415	.7332	.9164	1.0997
α = 55°	.1901	.2376	.2851	.3326	.3801	.4752	.5702	.6653	.7603	.9504	1.1404
α = 60°	.1972	.2465	.2958	.3450	.3943	.4929	.5915	.6901	.7887	.9858	1.1830

Table 2. — Values of L_1 (Figs. 119, 120, 122, 123, 125 and 131)

VALUES OF L AND L_1 FOR DOVETAILS AND V'S OF VARIOUS ANGLES WHERE D = DIAMETER OF WIRE

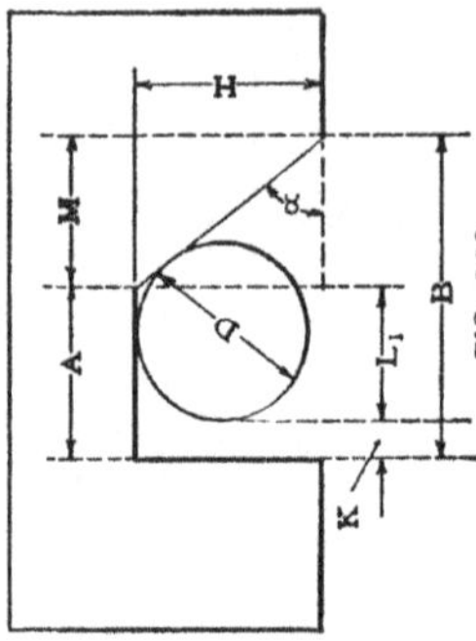

FIG. 123

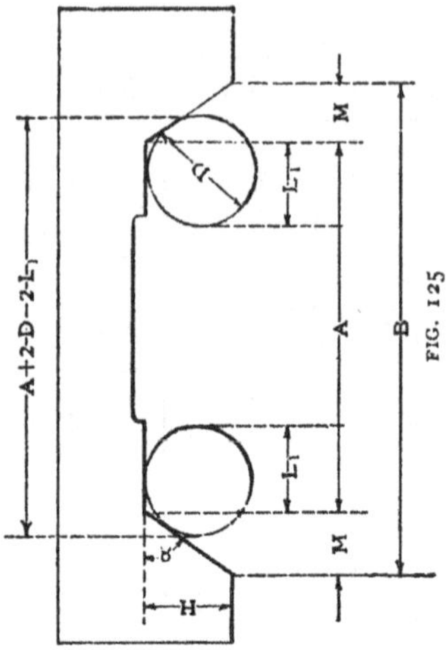

FIG. 125

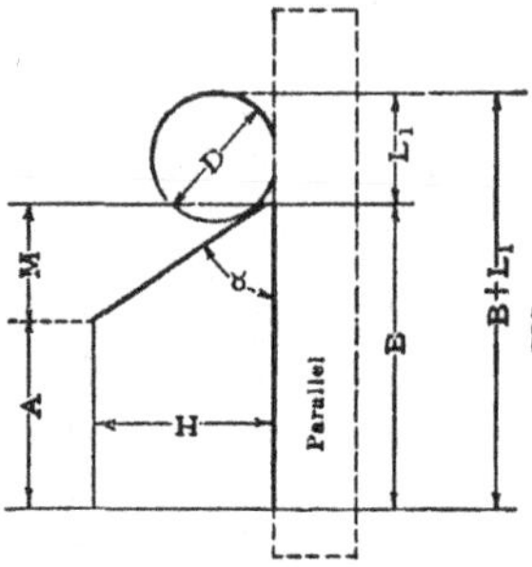

FIG. 122

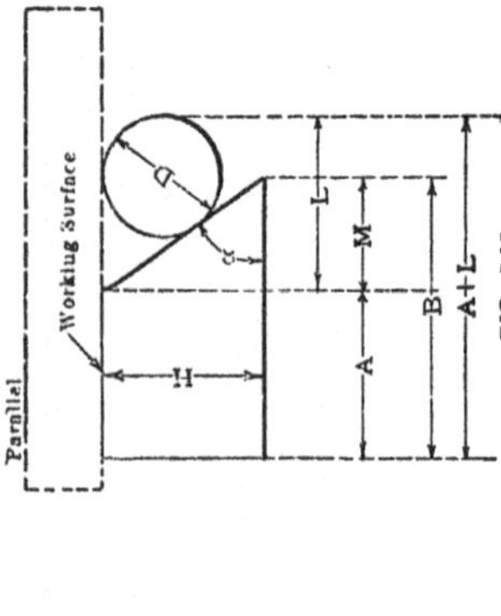

FIG. 121

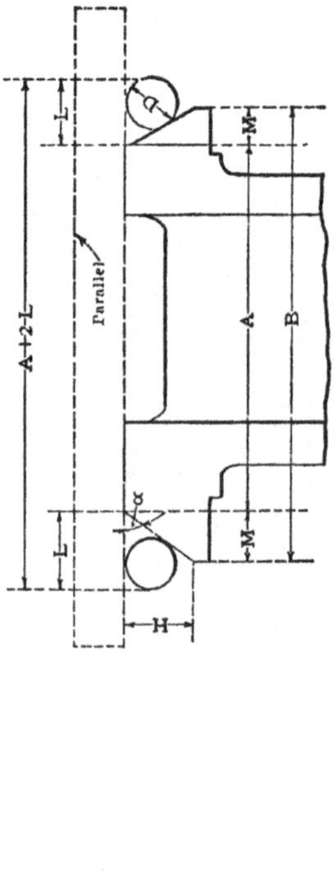

FIG. 124

MEASURING DOVETAILS

Table 2, which gives the value for L_1. Subtract this from A and the remainder is K. Another form of dovetail, used by machine-tool builders, is illustrated in Figs. 124 and 125. Fig. 124 represents the male and Fig. 125 the female slide. For Fig. 124, Table 1 has to be used, and for Fig. 125, Table 2. The rest will be understood from the illustrations.

MACHINE TOOL V'S

Last of all is to be considered the shape of the V, as illustrated in Figs. 126 and 127, used on lathes, planers, grinders and many other machine tools. It is advisable to plane the female part, Fig. 127, first, and fit the male part, Fig. 126, to it. The female part can be easily measured as shown later on the gage; but this operation on the male part is more difficult. A parallel has to be clamped on top of the V, as shown in Fig. 126, and dimensions $A+B+2L$ or $A+2L$ can be measured. A and B have to be given on the drawing, and for dimension L see Table 1. In manufacturing V's three gages will be needed, a roughing, a male and a female gage as illustrated in Figs. 128 to 131, respectively.

THE GAGES REQUIRED

The roughing gage, Fig. 128, will be used for determining the center distance of the two V's; tool-setting spots have been provided on top, to simplify the planing of the other side of the V's. To plane the V's of the roughing gage, the dimension required is $B+D$. The angle of the V has no influence on the measurements in this case, which is simply the center distance of the V's plus the diameter of the wire. The corners have been broken on the bottom as indicated in the sketch, to serve as a roughing gage for the female part also.

The finishing gage, Fig. 129, requires the dimensions $B+D$ (like the roughing gage) and L_2. Dimension $B+D$ controls the center distance of the V's, and L_2 the width N, which should be given on the drawing. The measurement L_2 can be taken in two different ways, as shown in Figs. 129 and 130, according to the measuring outfit at hand.

CONVENIENT TABLES

Tables 3 and 4 have been worked out for dimension L_2, considering different widths of V's, different diameters of wires and angles of 90 degrees and 120 degrees, using the formula:

$$L_2 = \frac{\frac{D}{2}}{sin.\ \frac{a}{2}} + \frac{D}{2} - \left(\frac{N}{2}\ cot.\ \frac{a}{2}\right).$$

The dimensions used for the measuring principle shown in Fig. 130 are

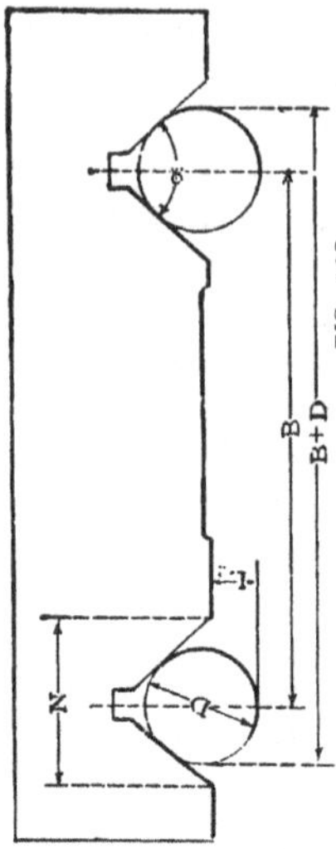

FIG. 129

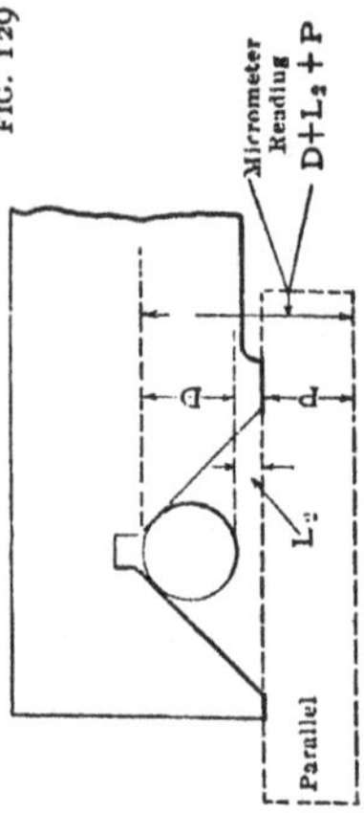

FIG. 130

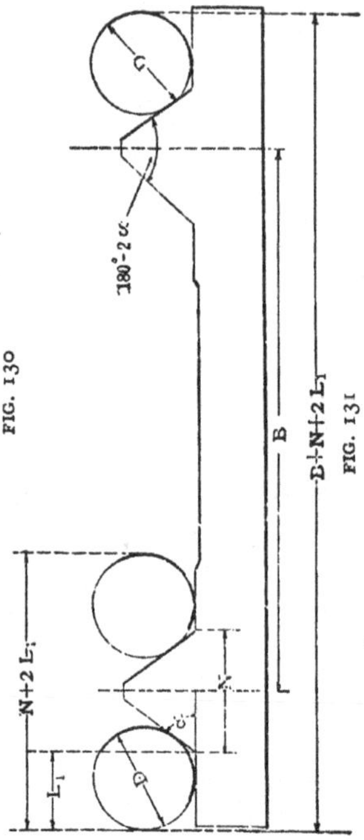

FIG. 131

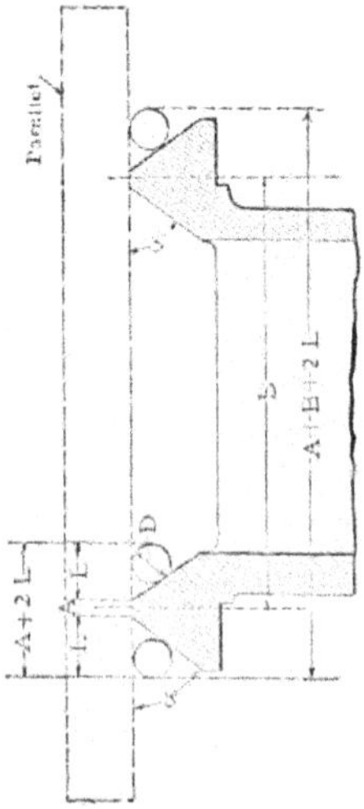

FIG. 126

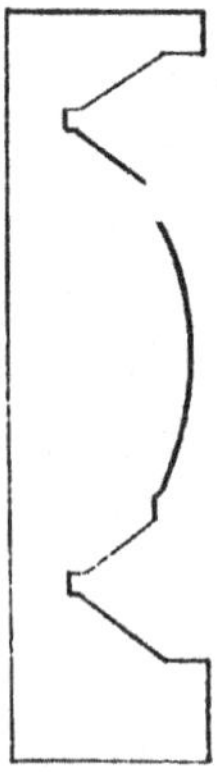

FIG. 127

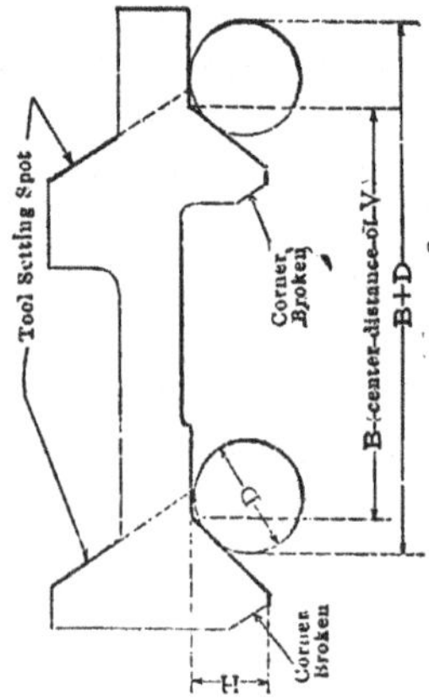

FIG. 128

MEASURING V'S

Dia. Wire = D	Width of V = N =	3/8	7/16	1/2	9/16	5/8	11/16	3/4	7/8	1	1 1/8	1 1/4	1 3/8
3/8"	$L_2=$	.2652	.2339	.2027	.1714	.1402	.1089	.0777	.0152	.0473	.1098	.1723	.2348
1/2"	$L_2=$	—	—	.3536	.3223	.2911	.2598	.2286	.1661	.1036	.0411	.0215	.0840
5/8"	$L_2=$	—	—	—	—	.4419	.4107	.3794	.3169	.2544	.1919	.1294	.0669
3/4"	$L_2=$	—	—	—	—	—	—	.5303	.4678	.4053	.3428	.2803	.2178
1"	$L_2=$	—	—	—	—	—	—	—	—	.7071	.6446	.5821	.5196
1 1/4"	$L_2=$	—	—	—	—	—	—	—	—	—	—	.8839	.8214
1 1/2"	$L_2=$	—	—	—	—	—	—	—	—	—	—	—	—
2"	$L_2=$	—	—	—	—	—	—	—	—	—	—	—	—

Dia. Wire = D	Width of V = N =	1 1/2	1 3/4	2	2 1/4	2 1/2	2 3/4	3	3 1/4	3 1/2	3 3/4	4
3/8"	$L_2=$	.2973	—	—	—	—	—	—	—	—	—	—
1/2"	$L_2=$	.1465	.2715	.3965	—	—	—	—	—	—	—	—
5/8"	$L_2=$	.0044	.1206	.2456	.3706	.4956	—	—	—	—	—	—
3/4"	$L_2=$	.1553	.0303	.0947	.2197	.3447	.4697	.5947	—	—	—	—
1"	$L_2=$	.4571	.3321	.2071	.0821	.0429	.1679	.2929	.4179	.5429	—	—
1 1/4"	$L_2=$	.7589	.6339	.5089	.3839	.2589	.1339	.0089	.1161	.2411	.3661	.4911
1 1/2"	$L_2=$	1.0607	.9357	.8107	.6857	.5607	.4357	.3107	.1857	.0607	.0643	.1893
2"	$L_2=$	—	—	1.4142	1.2892	1.1642	1.0392	.9142	.7892	.6642	.5392	.4142

TABLE 3. — VALUES OF L_2 (FIGS. 129 AND 130) FOR 90 DEGREE V'S WHERE N = WIDTH OF V AND D = DIAMETER OF WIRE

Dia. Wire = D	Width of V = N =	3/8	7/16	1/2	9/16	5/8	11/16	3/4	7/8	1	1 1/8	1 1/4	1 3/8	1 1/2	1 3/4	2
3/8"	$L_2=$	.2958	.2777	.2597	.2416	.2236	.2055	.1875	.1514	.1153	.0793	.0432	.0071	°.0290	°.1012	°.1734
1/2"	$L_2=$	—	—	.3943	.3763	.3583	.3402	.3222	.2861	.2500	.2139	.1778	.1417	.1057	.0335	°.0387
5/8"	$L_2=$	—	—	—	—	.4929	.4749	.4568	.4208	.3847	.3486	.3125	.2764	.2403	.1682	.0960
3/4"	$L_2=$	—	—	—	—	—	—	.5915	.5554	.5193	.4833	.4472	.4111	.3750	.3028	.2307
1"	$L_2=$	—	—	—	—	—	—	—	—	.7887	.7526	.7165	.6804	.6443	.5722	.5000
1 1/4"	$L_2=$	—	—	—	—	—	—	—	—	—	—	.9859	.9498	.9137	.8415	.7693
1 1/2"	$L_2=$	—	—	—	—	—	—	—	—	—	—	—	—	1.1830	1.1108	1.0387
2"	$L_2=$	—	—	—	—	—	—	—	—	—	—	—	—	—	—	1.5774

Dia. Wire = D	Width of V = N =	2 1/4	2 1/2	2 3/4	3	3 1/4	3 1/2	3 3/4	4	4 1/2	5	5 1/2	6	6 1/2	7
3/8"	$L_2=$	°.2455	°.3177	—	—	—	—	—	—	—	—	—	—	—	—
1/2"	$L_2=$	°.1108	°.1830	°.2552	°.3273	—	—	—	—	—	—	—	—	—	—
5/8"	$L_2=$	.0238	°.0484	°.1205	°.1927	°.2649	°.3370	°.4092	°.4814	—	—	—	—	—	—
3/4"	$L_2=$	.1585	.0863	.0142	°.0580	°.1302	°.2024	°.2745	°.3467	°.4910	°.6354	—	—	—	—
1"	$L_2=$	.4278	.3557	.2835	.2113	.1392	.0670	°.0052	°.0774	°.2217	°.3660	°.5104	°.6547	—	—
1 1/4"	$L_2=$	.6972	.6250	.5528	.4807	.4085	.3363	.2642	.1920	.0477	°.0967	°.2410	°.3854	°.5297	°.6740
1 1/2"	$L_2=$	.9665	.8943	.8222	.7500	.6778	.6057	.5335	.4613	.3170	.1727	.0283	°.1160	°.2604	°.4047
2"	$L_2=$	1.5052	1.4330	1.3609	1.2887	1.2165	1.1443	1.0722	1.000	.8557	.7113	.5670	.4227	.2783	.1340

TABLE 4. — VALUES OF L_2 (FIGS. 129 AND 130) FOR 120 DEGREE V'S WHERE N = WIDTH OF V AND D = DIAMETER OF WIRE

marked on the table with a circle (o). After the wire reaches a size where it strikes the parallel, the method of Fig. 129 has to be applied. For Fig. 130 the formula has to be altered and reads:

$$L_2 = \frac{N}{2} cot. \frac{a}{2} - \left[\frac{\frac{D}{2}}{sin. \frac{a}{2}} + \frac{D}{2} \right].$$

The finishing gage for the female slide, as illustrated in Fig. 131, requires dimension L_1, which is given under Table 2. As stated above, the female gage should be fitted to the male gage, and for this reason the table has not been figured for 30 degrees (or for a V of 120 degrees) as it will be needed very seldom.

CHAPTER XII

A Gage for Producing Accurate Tapers[1]

It has been customary in machine-tool shops, to produce a set of standard tapers to be used in connection with the line of manufacture. These standard taper gages are carefully kept in the company's safe, and are used for reference only. Another set of gages, an exact copy of the first set, is then made for shop use, and this second set is occasionally compared with the first. This method seems somewhat cumbersome, and to overcome it the tool illustrated in Fig. 132 was designed and made some years ago.

It is very evident that if two round disks of unequal diameter are placed on a surface plate a certain distance apart, two straight-edges touching these two disks will represent a certain taper. It is also evident that with the measuring instruments now in use it is a very simple matter to measure accurately the diameters of the two disks, and also to measure accurately the distance these disks are apart. It will readily be seen that these three dimensions accurately and positively determine the taper represented by the straight-edges touching the rolls. If now a proper record is made of these three dimensions it is very evident that these conditions can be reproduced, and this is in short the idea which underlies the use of the tool herewith illustrated.

Some years ago a special taper plug was made and sent to one of the manufacturers of arms in Russia. Upon the plug being received, the management of the factory requested that a female gage to fit the plug be forwarded. On the plan of the tool illustrated the plug was reproduced from the dimensions noted at the time. A second plug was made, the female gage was produced and shipped. When the female gage was received, it fitted perfectly with the plug previously shipped, illustrating the ease with which tapers could be reproduced by means of the tool described.

The following formulas may be of service in connection with a gage of this character:

TO FIND CENTER DISTANCE BETWEEN DISKS

Suppose we have two disks as shown in Fig. 133, whose diameters are respectively $1\frac{1}{4}$ and 1 inch. We wish to construct a taper of $\frac{3}{4}$ to the foot and must determine the center distance l between disks in order that the gage jaws when touching both disks shall give that taper.

[1] Contributed to the *American Machinist* by C. C. Stutz.

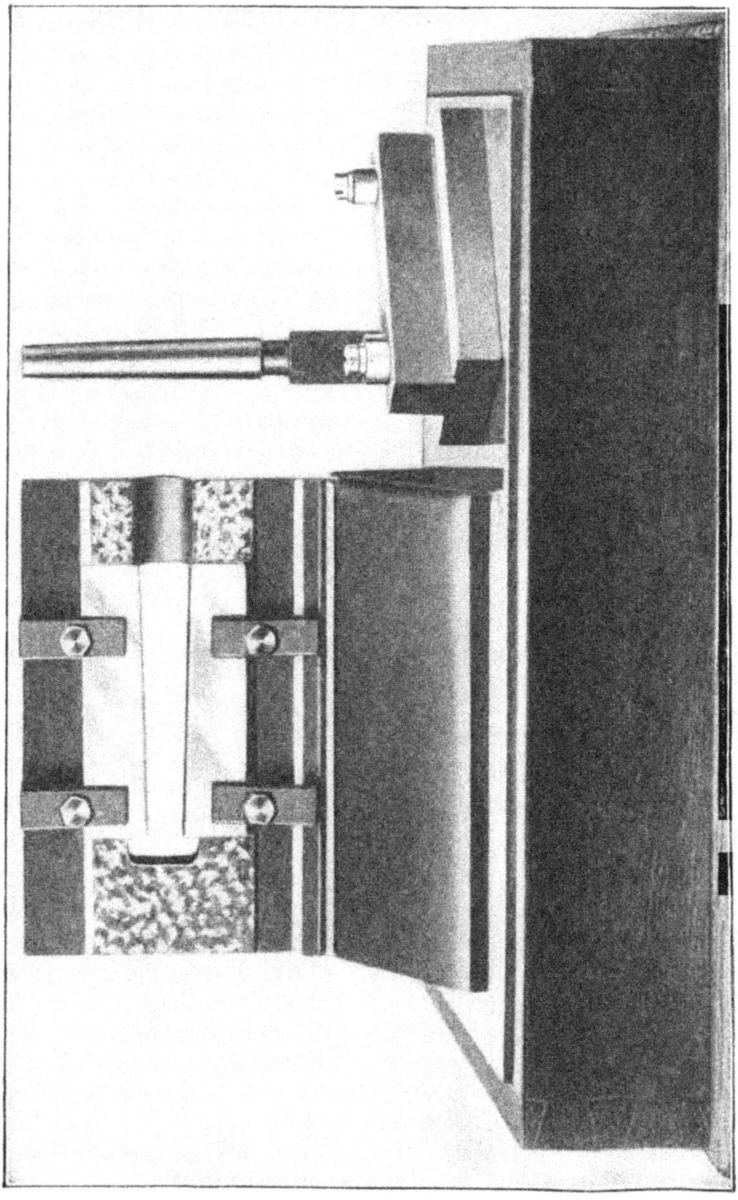

FIG. 132. — TAPER GAGE WITH ADJUSTABLE JAWS AND SETTING DISKS

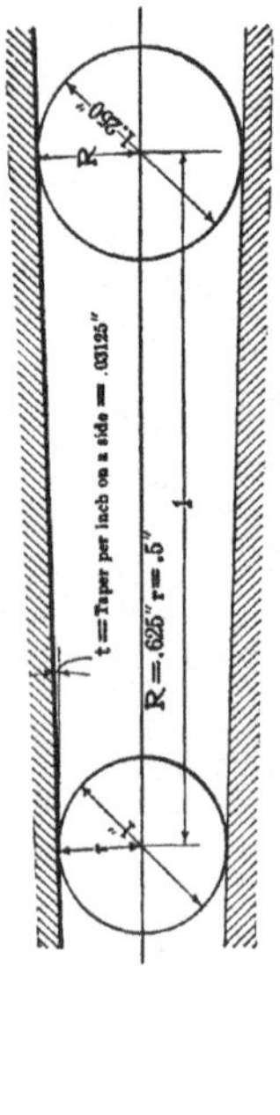

Taper per foot = ¾"

FIG. 133

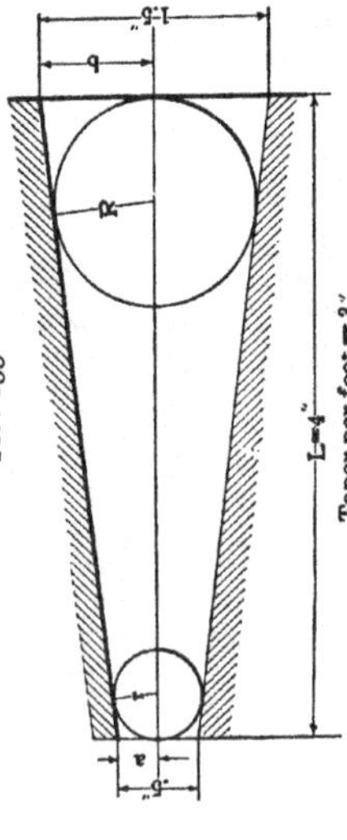

Taper per foot = 3"

FIG. 134

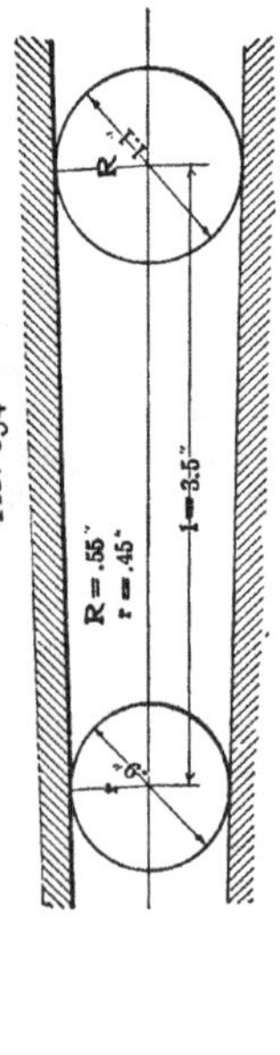

FIG. 135

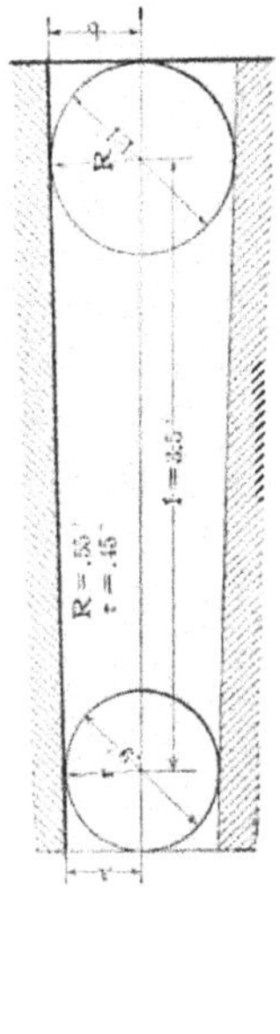

FIG. 136

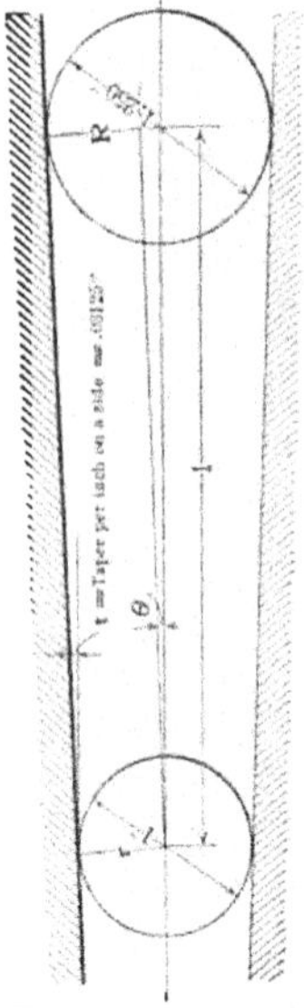

FIG. 137

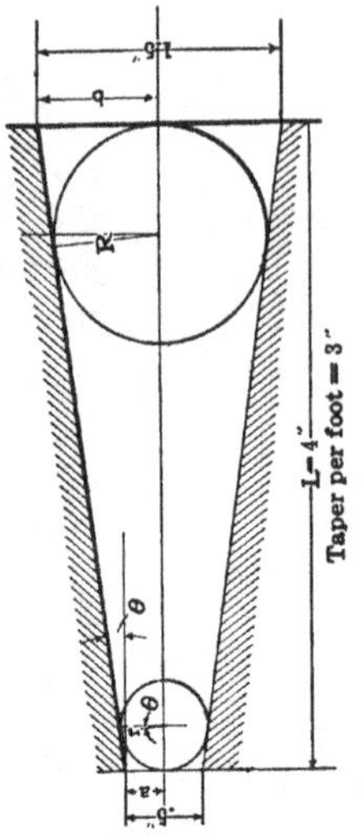

Taper per foot = 3"

FIG. 138

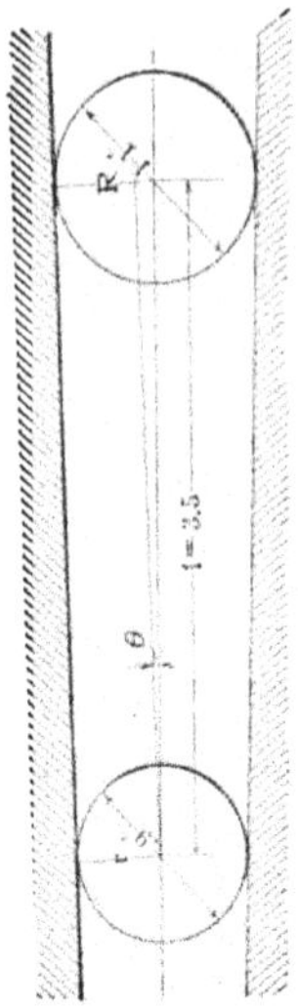

FIG. 139

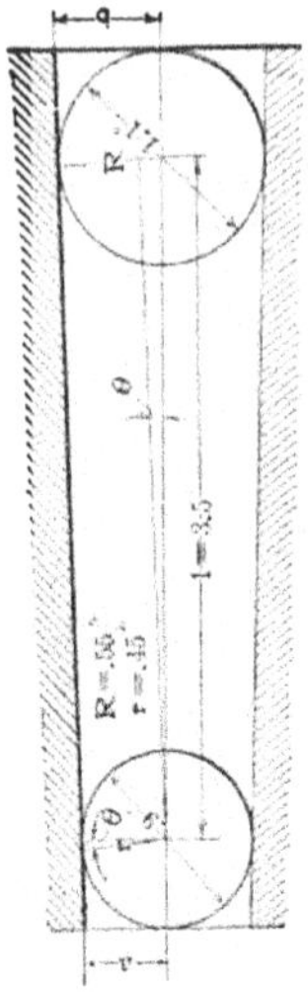

FIG. 140

To find Center Distance (l), refer to Fig. 133.

$$l = \frac{R - r}{t}\sqrt{1 + t^2}.$$

To find Disk Diameters, refer to Fig. 134.

$$r = \frac{a}{L}\left[\sqrt{L^2 + (b - a)^2} + (b - a)\right].$$

Dia. Small Disk = 2 r

$$R = \frac{b}{L}\left[\sqrt{L^2 + (b - a)^2} - (b - a)\right].$$

Dia. Large Disk = 2 R

To find Taper Per Foot (T), refer to Fig. 135.

$$T = 24\left(\frac{R - r}{\sqrt{l^2 - (R - r)^2}}\right).$$

To find Width of Opening at Ends, refer to Fig. 136.

$$a = r\left[\frac{l}{\sqrt{l^2 - (R - r)^2}} - \sqrt{\left(\frac{l}{\sqrt{l^2 - (R - r)^2}}\right)^2 - 1}\right].$$

Width of Opening at Small End = 2 a.

$$b = R\left[\frac{l}{\sqrt{l^2 - (R - r)^2}} + \sqrt{\left(\frac{l}{\sqrt{l^2 - (R - r)^2}}\right)^2 - 1}\right].$$

Width of Opening at Large End = 2 b.

ALGEBRAIC FORMULAS FOR TAPER GAGE

To find Center Distance (l), refer to Fig. 137.

$l = \frac{R - r}{Sin\ \theta}$, in which θ = the angle whose tangent is the taper per inch on a side, or t.

To find Disk Diameters, refer to Fig. 138.

$r = \frac{a}{Cos\ \theta}(1 + Sin\ \theta)$, $R = \frac{b}{Cos\ \theta}(1 - Sin\ \theta)$, in which θ = the angle whose tangent $= \dfrac{\frac{b - a}{2}}{L}$.

Dia. Small Disk = 2 r.

Dia. Large Disk = 2 R.

To find Taper Per Foot (T), refer to Fig. 139.

$T = 24\,(tan\,\theta)$, in which θ = the angle whose sine $= \frac{R - r}{l}$.

To find Width of Opening at Ends, refer to Fig. 140.

$a = r\ tan\left(45° - \frac{\theta}{2}\right)$, $b = R\ tan\left(45° + \frac{\theta}{2}\right)$, in which θ = the angle whose sine $= \frac{R - r}{l}$.

Opening at Small End = 2 a.

Opening at Large End = 2 b.

TRIGONOMETRICAL FORMULAS FOR TAPER GAGE

Let R = radius of large disk, or 0.625 inch.
r = radius of small disk, or 0.500 inch.
t = taper per inch on side, or

$$\frac{0.750}{24} = 0.03125 \text{ inch.}$$

Then

$$l = \frac{R - r}{t}\sqrt{1 + t^2} = \frac{0.125}{0.03125}\sqrt{1.000976} = 4 \times 1.0005 = 4.002 \text{ inches.}$$

TO FIND DISK DIAMETERS

Suppose the gage jaws are to be set as in Fig. 134 for a 3-inch per foot taper whose length is to be 4 inches. The small end is to be exactly ½ inch and the large end for this taper will, therefore, be 1½ inches. What diameter must the disks be made so that when the jaws are in contact with them and the distance L over the disks measure 4 inches, the taper will be exactly 3 inches per foot? Here a represents one-half the width of opening at the small end, and b one-half the width of opening at the large end. The radius of the small disk may be found by the formula:

$$r = \frac{a}{L}\left\{\sqrt{L^2 + (b - a)^2} + (b + a)\right\}.$$

Then

$$r = \frac{0.250}{4}\left(\sqrt{16 + 0.25} + 0.5\right) = 0.0625(4.0311 + 0.5) = 0.2832.$$

Diameter small disk = 0.2832 inch × 2 = 0.5664 inch.
For the large disk:

$$R = \frac{b}{L}\left\{\sqrt{L^2 + (b - a)^2} - (b - a)\right\}.$$

Then

$$R = \frac{0.75}{4}\left(\sqrt{16 + 0.25} - 0.5\right) = 0.1875(4.0311 - 0.5) = 0.6621.$$

Diameter large disk = 0.6621 inch × 2 = 1.3242 inches.

TO FIND TAPER PER FOOT

In duplicating a taper the gage jaws may be set to the model, and by placing between the jaws a pair of disks whose diameters are known the taper per foot may be readily found. For example, the jaws in Fig. 135 are set to a certain model, two disks 0.9 and 1.1 inch diameter are placed between them and the distance over the disks measured, from which dimension l (which is 3.5 inches) is readily found by subtracting half the diameters of the disks. Here l represents the center distance as in Fig.

133. To determine the taper per foot which may be represented by T, the formula is:

$$T = 24\left(\frac{R - r}{\sqrt{l^2 - (R - r)^2}}\right).$$

Then

$$T = 24\left(\frac{0.1}{\sqrt{12.25 - 0.01}}\right) = 24\left(\frac{0.1}{3.4985}\right) = 0.684.$$

Taper per foot = 0.684 inch.

TO FIND WIDTH OF OPENING AT ENDS

If, with the ends of the gage jaws flush with a line tangent to the disk peripheries as in Fig. 136, we wish to find the width of the opening at the small end where a represents one-half that width, the following formula may be of service, the disks being as in the last example 0.9 and 1.1 inch diameter respectively, and the center distance 3.5 inches:

$$a = r\left\{\frac{l}{\sqrt{l^2 - (R - r)^2}} - \sqrt{\left(\frac{l}{\sqrt{l^2 - (R - r)^2}}\right)^2 - 1}\right\}.$$

Then

$$a = 0.45\left\{\frac{3.5}{\sqrt{12.24}} - \sqrt{\left(\frac{3.5}{\sqrt{12.24}}\right) - 1}\right\} = 0.45\,(1.00043 - \sqrt{0.00086}) =$$

$$0.45\,(1.00043 - 0.0293) = 0.437.$$

0.437 inch × 2 = 0.874 inch, width of opening at small end of gage.

Similarly the width of opening at the large end of the gage may be found as follows, where b = half the width at large end.

$$b = R\left\{\frac{l}{\sqrt{l^2 - (R - r)^2}} + \sqrt{\left(\frac{l}{\sqrt{l^2 - (R - r)^2}}\right)^2 - 1}\right\}$$

Then

$$b = 0.55\left\{\frac{3.5}{\sqrt{12.24}} + \sqrt{\left(\frac{3.5}{\sqrt{12.24}}\right)^2 - 1}\right\} = 0.55\,(1.00043 + \sqrt{0.00086}) =$$

$$0.55\,(1.00043 + 0.0293) = 0.5663.$$

0.5663 inch × 2 = 1.1326 inch = width of gage opening at large end.

The formulas for a and b appear a little complicated; actually, however, they are simple enough. The expression

$$\frac{l}{\sqrt{l^2 - (R - r)^2}}$$

which appears twice in each formula is readily given its numerical value and upon this substitution the appearance is generally simplified.

FINDING CENTER DISTANCE BY TRIGONOMETRY

Some may prefer to solve the foregoing problems by trigonometry. The diagram Fig. 137 duplicates the conditions in Fig. 133, and it is obvious that t, the taper per inch on each side (0.03125") $= tan.\ \theta$, which angle we find to be 1° 47′ 20″. To find center distance l the following formula may be used:

$$l = \frac{R - r}{sin.\ \theta}.$$

Then

$$l = \frac{0.625 - 0.5}{sin.\ 1° 47' 20''} = \frac{0.125}{0.03123} = 4.002 \text{ inches.}$$

DISK DIAMETERS

In Fig. 138 the conditions of the problem Fig. 134 are reproduced, only here r and R are to be found by trigonometry in determining the diameters of the two disks. Angle θ is readily found as the tangent of this angle is obviously equal to

$$\frac{\frac{1.5 - 0.5}{2}}{4} = 0.125 = tan.\ 7° 8'.$$

To find r use the formula

$$r = \frac{a}{cos.\ \theta}(1 + sin.\ \theta).$$

Then

$$r = \frac{0.25}{0.99226}(1 + 0.12418) = 0.2519 \times 1.12418 = 0.2832.$$

Diameter small disk = 0.2832 inch × 2 = 0.5664 inch.
Similarly,

$$R = \frac{b}{cos.\ \theta}(1 - sin.\ \theta).$$

Then

$$R = \frac{0.75}{0.99226}(1 - 0.12418) = 0.7549 \times 0.87582 = 0.6621.$$

Diameter large disk = 0.6621 inch × 2 = 1.3242 inch.

TAPER PER FOOT

In Fig. 139 we have the problem reproduced from Fig. 135, of finding the taper per foot when the disk diameters and their distance apart are known. To solve this by trigonometry we may consider angle θ as representing the slope on one side of the taper, hence the total taper per foot or T may be expressed as $T = 24\ (tan.\ \theta)$. We can readily find this angle as follows:

$$Sin.\ \theta = \frac{R - r}{l} = \frac{0.55 - 0.45}{3.5} = \frac{0.10}{3.5} = 0.0285 = sin.\ 1° \ 38'.$$

Then $T = 24\ (tan.\ \theta) = 24\ (tan.\ 1° \ 38') = 24\ (0.02851) = 0.6842$. Taper per foot = 0.6842 inch.

WIDTH OF OPENING

Considering now the case in Fig. 140, where, as in Fig. 136, it is desired to find the width of opening in the gage at a point tangential to the extremities of the disks, we may solve a by the formula

$$a = r\ tan.\left(45° - \frac{\theta}{2}\right).$$

From the data of the problem we easily find the angle θ; for $sin.\ \theta = \frac{R - r}{3.5} = 0.0285 = sin.\ 1° \ 38'$ as in the preceding problem.

Applying formula, $a = r\ tan.\left(45° - \frac{\theta}{2}\right)$ we have

$$a = 0.45\ (tan.\ 44° \ 11') = 0.45\ (0.97189) = 0.437.$$

0.437 inch $\times$ 2 = 0.874 inch width of opening at small end of gage. To find dimension b, Fig. 140, by trigonometry we use the formula

$$b = R\ tan.\left(45° + \frac{\theta}{2}\right) = 0.55\ (tan.\ 45° \ 49') = 0.55\ (1.02892) = 0.566.$$

0.566 inch $\times$ 2 = 1.132 inch, width of gage opening at large end.

CHAPTER XIII

The Microscope in the Tool Room

Of the many tools and appliances adapted to facilitate the production of accurate work in the tool room, one instrument at least has been sadly neglected. This is the microscope, which, although commonly regarded as a scientific instrument whose field of usefulness is confined to the laboratory, has, nevertheless, a great number of applications in connection with tool work. It will be understood that reference is made here not to the ordinary magnifying glass, but to the compound microscope with cross hairs in the eye piece, an instrument whose adaptibility to the purposes of the toolmaker will be outlined here in connection with the accompanying half-tones and line engravings.

Several years ago Walter Gribben described in the columns of the *American Machinist* a method of measuring a steel tape and a scheme for centering work on the lathe face plate, using the microscope in both cases for setting the work accurately. A few others have since then adopted this device for use on some classes of work; so far as is known, however, not over three or four machine shops at the most have made any extensive use of this convenient instrument on general tool work. One of the believers in and users of the microscope for many operations is W. A. Warman, of the Keller Mechanical Engraving Company, of New York City.

Mr. Gribben's instrument is shown mounted on a surface plate in Fig. 141, and attached to the slide rest of a bench lathe in Fig. 142. Fig. 143 illustrates a form of microscope used by Mr. Warman and manufactured by the Keller company, this instrument being mounted in a very similar fashion to a regular surface gage. In the cases of both instruments shown, the working distance is about 1.4 inches and the magnifying power about 25 diameters. The cross hairs intersect at an angle of 30 degrees, as indicated in the numerous sketches reproduced herewith.

LOCATING WORK ON THE FACE PLATE

One of the most convenient uses to which the microscope can be put is in locating a job, say a jig plate, by a pair of lines scribed on the surface, there being no necessity in this case of center punching the point of intersection and using this as a center by which to indicate the plate.

In Fig. 144 let diagram *A* indicate the field of view of the microscope, showing the cross hairs, while *B* shows in the field two lines (of course greatly magnified) crossing at a point where it is desired to bore a hole in the plate. Prior to attempting to center the work, the microscope itself is centered by focusing it upon some point on the plate which appears to be stationary while the lathe is running, and also rotating the microscope tube itself in its clamping sleeve until one of the cross hairs is approxi-

FIG. 141. — THE MICROSCOPE ON THE SURFACE PLATE

mately horizontal. With the microscope shown in Fig. 142 any slight vertical adjustment necessary is easily obtained, owing to the instrument being held by friction in the clamp which is attached to the slide rest of the lathe. Thus the tube may be tilted slightly up or down and will remain stationary wherever set. It will be understood, of course, that a slight inclination in the axis of the microscope does not in any way affect the correctness of the results obtained.

FIG. 142. — THE MICROSCOPE ON THE BENCH LATHE

Now assuming that we have the cross hairs approximately central with the axis of the lathe spindle, the lines on the work, appearing as at *B*, Fig. 144, the next operation is to shift the work to bring the horizontal line approximately central with the horizontal cross hair as shown at *C*. While this move apparently means lifting the work slightly on the plate, actually it means moving it downward, as the compound microscope

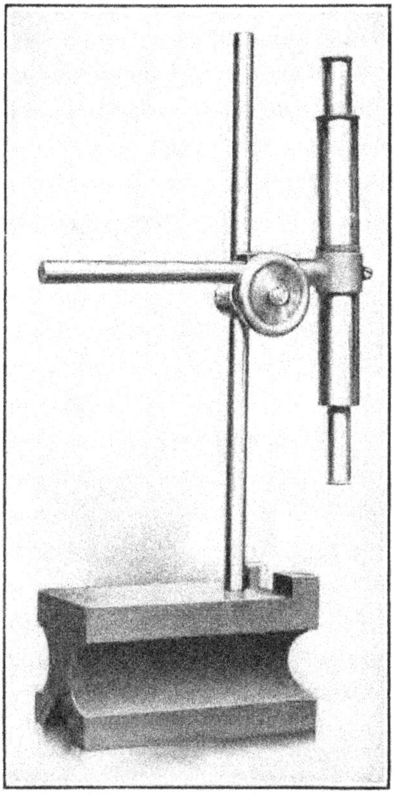

FIG. 143. — A HANDY MICROSCOPE

reverses all images in the field. It is, therefore, easier to start out on the centering operation with the work in reality just slightly above center, allowing it to be shifted downward across the plate. After moving the work until the horizontal line appears to be somewhere near central with the horizontal cross hair of the microscope, as shown at *C*, the microscope itself is moved until the horizontal hair is just on the edge of the work line, as at *D*. The amount which the instrument is shifted is indicated (greatly magnified) at *a*. Next, the face plate is turned half over, and then, assuming that the lines scribed on the plate appear as at *E*, the amount which the work is apparently out of center is indicated doubled by the gap *b*. Now the instrument is shifted slightly, or until the horizontal cross hair has apparently split the previous error, shown at *c*, its position then being as indicated at *F*, and the work is shifted to bring the edge of the line again coincident with the cross hair, as at *G*. The face plate and work are again turned half-way over, and if the setting is correct, the other side of the line will coincide with the cross hair, as shown at *H*. If there is still an error in the setting, the microscope is

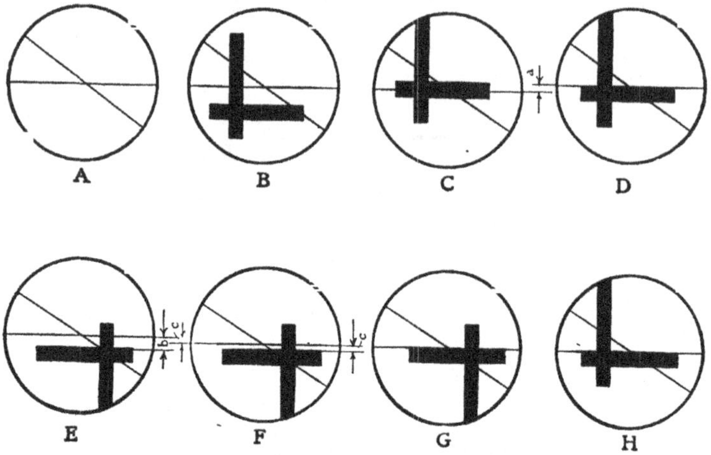

FIG. 144. — CENTERING WORK ON THE FACE PLATE

again shifted to take up half this error and the work shifted the other half. The object, of course, in turning the face plate half-way round is to bring opposite sides of the line in turn against the microscope cross hair, thus demonstrating that both edges of the work line are equidistant from the center line of the lathe spindle. The work, now being central in one direction, is adjusted until the other line has been centered with the cross hair; a final sight may then be taken on the line first centered, to determine if readjustment by this line is necessary.

While the foregoing paragraph and the sketches in Fig. 144 show the method of centering one line first and repeating the series of operations for the other line, actually the work may be expedited by approximately centering each work line before attempting to center either accurately by the edges, as shown at *D, E, F, G* and H. While this method may appear to be somewhat lengthy, it is in fact a very quick way of centering a job. Its one great advantage is that the indicating point, that is, the intersection of the cross hairs, by which the work is centered, is really out in the air; that is, there is no mechanism to be brought into contact with the face of the work. Therefore the scribed lines are sufficient for centering purposes, thus doing away with the necessity for putting a center-punch mark at the point of intersection, which invariably means an error, greater or less according to the skill of the man who handles the punch.

Where care is taken to scribe lines on the work so fine that they are invisible to the naked eye, or nearly so, the work may be centered directly by adjusting until the lines show central with the cross hairs in the instrument, thus avoiding testing first one edge of a line and then the other. With carefully scribed lines the error in centering this way is exceedingly minute.

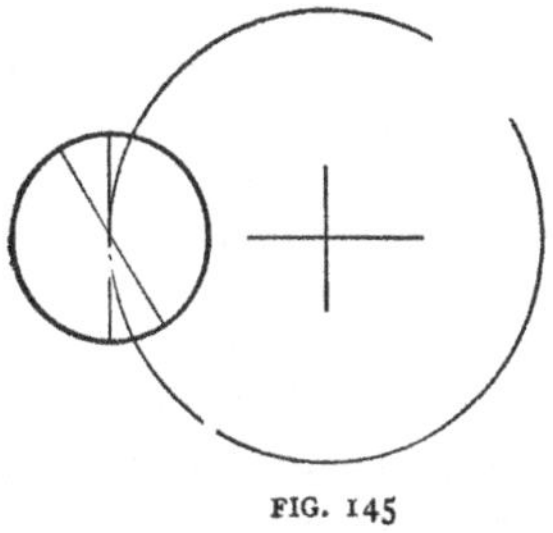

FIG. 145

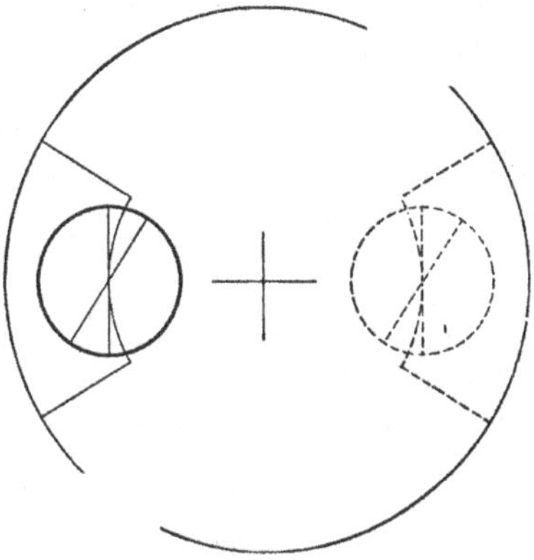

FIG. 146

FACE-PLATE WORK

Sometimes in boring a hole in a piece held on the face plate, there is a possibility of the work shifting, during the roughing-out process. As a means of guarding against this, a circle is sometimes scribed from the center-punch mark; as the hole is enlarged, this circle can be watched and any movement of the plate will, of course, be shown by the travel of this circle. When boring out a piece with the aid of the microscope for setting, the test circle may be scribed by means of a scribing block placed on the lathe rest, making a fine line, as indicated in Fig. 145. After roughing out the hole, the microscope may then be set with one cross hair tangent to the circle, and any out of center movement which the work may make during the boring of the hole will be readily detected by watching the travel of the circle past the vertical hair, and the work can be shifted until no apparent to and fro movement of this circle takes place in reference to the cross hair when the lathe spindle is slowly rotated.

MEASURING THE RADIUS OF A SEGMENT

While we are considering the use of the microscope in connection with face-plate work, one other example may be cited showing the convenience of this instrument. Suppose we have, as indicated in Fig. 146, a segment to be bored out to a given radius. When there are a number of such segments, they may be placed on the face plate, bored out and calipered precisely as an ordinary ring would be bored and measured; but where there is only one piece to machine, the proposition is an entirely different one. The usual expedient is to bolt on a piece of scrap iron opposite the segment being machined. With a microscope attached to the cross slide, we may test the correctness of the work by bringing the slide forward until the vertical hair is tangent to the arc bored out as shown in Fig. 146; the face plate may then be turned half over and the cross slide moved to the rear (counting the turns of the screw and reading the micrometer graduations) a total distance equal to twice the required radius of the work, or in other words equal to the diameter swept out. The microscope and the work in this second position are indicated by the dotted line, Fig. 146. If the piece is turned out to the right radius, the vertical hair will be tangent to the arc in its second position, just as it was when tested previously at the front of the center. With an accurate screw for operating the cross slide and the graduated collar thereon, we thus have a very handy means of calipering work, either external or internal.

COMPARING LATHE FEED SCREWS WITH A STANDARD SCALE

The half-tone, Fig. 142, shows a standard scale clamped on the face plate of the bench lathe. This represents a convenient means of testing the accuracy of the cross-feed screw with the scale, or *vice versa*, testing the scale against the screw. For a job of this kind where the lines have considerable width, it is probably better to focus the vertical hair on one side of the line, as indicated in Fig. 147. The microscope may then be moved transversely, by the cross-feed screw, bringing the vertical hair to the corresponding edge of the different lines; any discrepancies between the scale and the screw will, of course, be readily seen. Similarly the longitudinal-feed screw may be tested by placing a standard scale at the back of the bed and focusing the microscope thereon by mounting it in a vertical position.

INSPECTION OF THREADED WORK

Figs. 148, 149 and 150 show a way of testing the accuracy of taps or other threaded work by comparing with the lathe lead screw. A tap is

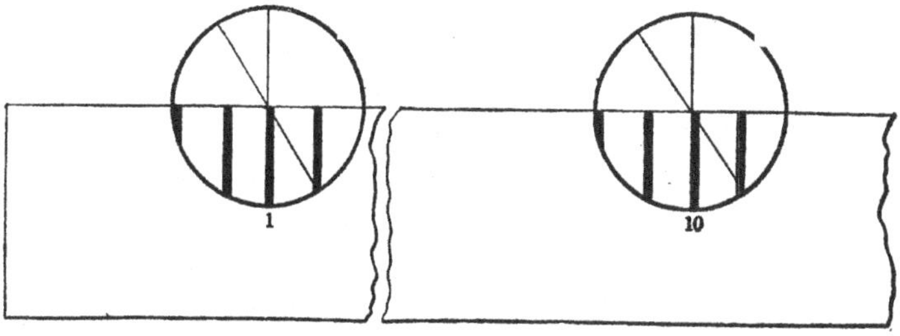

FIG. 147. — SHOWING CROSS HAIRS SIGHTED ON SCALE GRADUATIONS

FIG. 148. — INSPECTING A TAP

likely to change more or less during the hardening process, and it is oftentimes very desirable to know just what the actual change in the lead amounts to. Mr. Warman's instrument, in Fig. 148, is shown focused on the thread of the tap held between the lathe centers. The field and the portion of the work coming therein is indicated by the sketch, Fig. 149, where the microscope tube is twisted in its holder sufficiently to bring one hair into coincidence with the helix of the thread; although the top of the thread when magnified has a rounded appearance, there is little difficulty in setting the microscope cross hair central with the actual top,

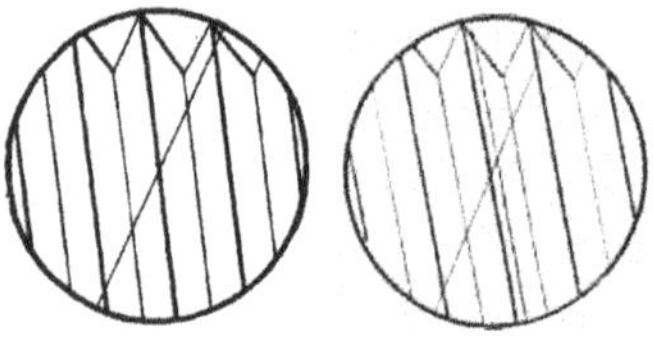

FIG. 149

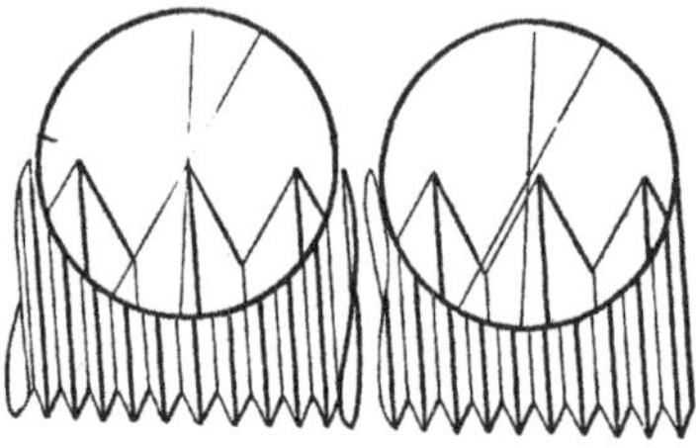

FIG. 150

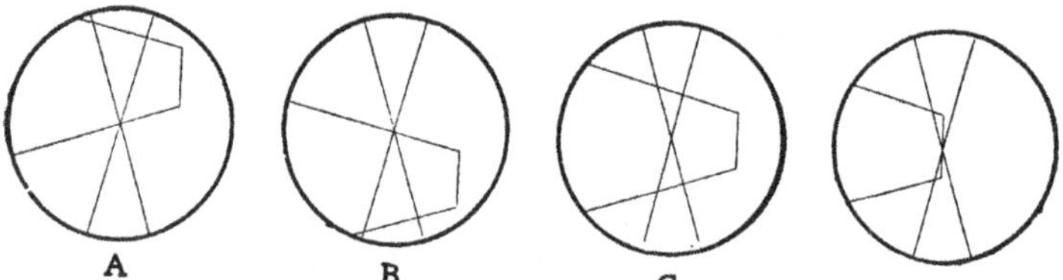

A B C D

FIG. 151

THREAD AND THREAD-TOOL INSPECTION

as a very small amount off center is easily detected. If the carriage is moved along by the lead screw, the cross hair as brought successively over the different threads will show any error in the work as compared with the lead screw. When the microscope is placed on the compound rest, having a micrometer dial on the screw, the actual distance from center to center of the threads may be tested; the total error between the two most widely separated threads on the job may also be found directly. Sometimes it is more convenient to set the microscope as suggested in Fig. 150, one cross hair in this case coinciding with the edge of the slope on the threads. In both sketches, Figs. 149 and 150, it will be seen that a slight error is indicated as detected by the second setting of the instrument.

A WORM-THREAD TOOL AND A FLY-CUTTER JOB

Fig. 151 shows an application of the microscope in grinding a worm-thread tool. The tool is first ground to the proper angle, that is, 14½ degrees with the sides of the shank, leaving the flat at the end too narrow. Then the tool is set in the chuck, the microscope placed vertically in the slide rest and brought to bear in succession, at points *A*, *B*, *C* and *D*. The longitudinal slide for this job is moved until the cross hairs, when focused on opposite edges of the tool, as at *A* and *B*, give the correct width at the pitch line for the tool, as read on the cross-slide micrometer collar. The microscope is then moved to the position indicated at *C*, that is, about the middle of the width of the tool at the pitch line, and by means of the longitudinal screw the distance to the end of the tool, as at *D*, is found. After the microscope has been adjusted, as at *D*, the actual length of the tool from the pitch line to the end is read off from the micrometer on the slide-rest screw, and from this distance may be subtracted the correct length from pitch line to end of tool (as found by calculation, or from a table), showing at once how much must be ground off the end of the tool to bring it to standard width.

Mr. Gribben, in making a fly cutter to cut a slot of a certain width, used the microscope as indicated in Fig. 152 to test the working width of the cutter. The latter was held in the lathe chuck, with the microscope held vertically in the slide rest, testing successively as indicated in the two views in Fig. 152, sighting first on the cutting edge of the cutter toward the front of the lathe, then turning the spindle and cutter half-way over and sighting on the same edge of the cutter turned toward the back of the lathe. The width of path which the cutter would describe in operation was therefore shown by the reading of the micrometer on the cross screw in adjusting the microscope from the first sighting position to the second.

FINDING THE DISTANCE BETWEEN LATHE-BED V'S

Some time ago Mr. Gribben wished to make a templet for his bench lathe ways and first naturally wished to find the center distance between the front and rear V's. For this purpose he made a shoe, shown in Fig. 153, which was cut out and scraped to fit the V's, as indicated. Striking a line on the upper face of the shoe and setting his microscope vertically on his slide rest, the shoe resting on the front V, he adjusted the rest until the point of intersection of the microscope cross hairs was over the line on the shoe, noting the reading of the graduations on the cross-slide screw at this point. The shoe was then shifted to the rear V, the slide rest run back and the microscope again sighted on the line of the shoe. The exact distance moved (which was the exact center distance) was of course

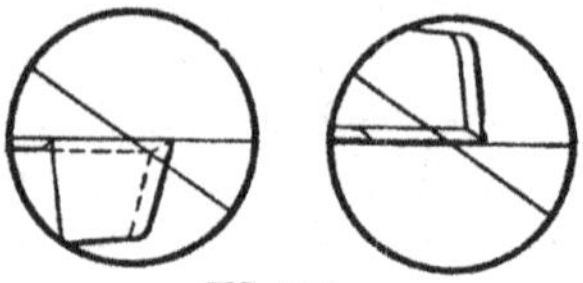

FIG. 152

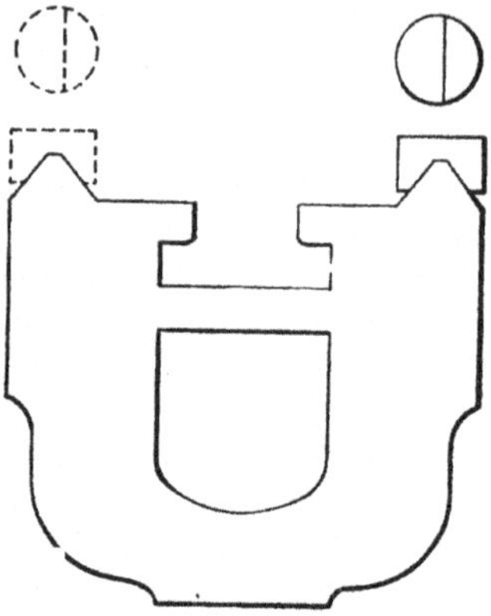

FIG. 153

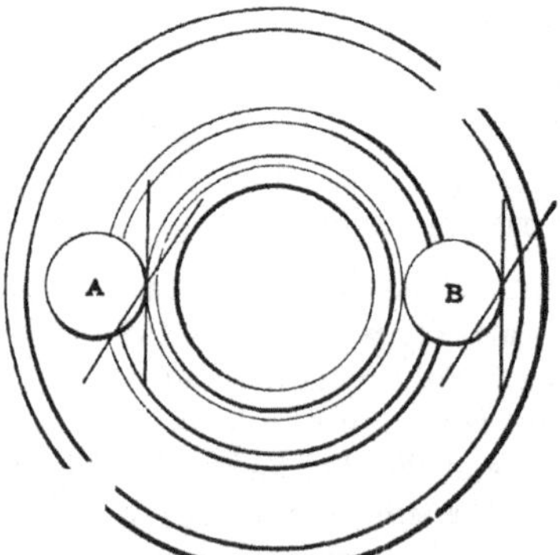

FIG. 154

OTHER OPERATIONS WITH THE MICROSCOPE

readily determined by keeping track of the number of turns and parts of a turn of the slide-rest screw.

A BALL-RACE MEASUREMENT

Fig. 154 shows another interesting scheme, namely, measuring the diameter of the race in a ball bearing which apparently was made too small for the number and size of balls used. The bearing worked improperly, and it was evident that the balls were being crowded. The minimum diameter permissible, that is, a diameter which would bring all the balls in contact with one another, was readily determined by calculation; then the actual diameter of the race was found by measuring with a microscope mounted vertically on the slide rest of the lathe, the ball race being clamped to the bed of the lathe in horizontal position and containing one ball only. First the ball was sighted, as at *A*, Fig. 154, with one cross hair tangent to the inner side of the ball; then the latter was rolled around to the opposite side of the race and the slide rest adjusted by the screw until the same hair was again tangent with the outside of the ball, as at *B*. The distance the microscope was moved was shown by the micrometer collar. This test showed that the race was actually 0.0016 too small in diameter, or that much less than the proper diameter as determined by calculation.

WORK ON THE SURFACE PLATE

By referring again to Fig. 141, some further idea of the convenience of the microscope for laying out and other surface-plate operations will be obtained. The disk clamped on the angle plate in this illustration has drawn upon its surface a line to which it is desired to mill or turn off the edge of the disk, for the purpose of leaving a flat edge at this point. It is desirable, therefore, to set this disk in such a position that the line shall be parallel to the bottom of the angle plate upon which the work will be carried while the operation of flatting off the edge is performed. It is obviously a very simple matter to focus the cross hairs at one end of the line, then slide the work to and fro past the field of the microscope, adjusting the disk until the line coincides throughout its length with the intersection of the cross hairs. This illustration will suggest a number of operations, in connection with tool-room work that come within the field of the microscope.

A SPECIAL MICROSCOPE FIXTURE

On the engraving machine built by the Keller Mechanical Engraving Company a four-cornered cutting tool is used, which is of rhombic section, so that two corners only do the cutting. The cutter is tapered through the greater part of its cutting length and rounded at the end, the amount of taper and the radius of the arc at the end varying with different classes and sizes of work. A corresponding taper and rounded end are given to the guide pin or tracer, which throughout the operation of mechanically engraving a die is held in contact with the model and operates a pivoted bar carrying the cutter for milling out the dies. The distance between the guide pin or tracer and the cutter is a variable quantity in accordance with the reduction required in the work, the model being made considerably larger than the die which is required. Consequently, the cutter itself is made of different diameter than the tracer, the difference of course varying with the amount of reduction required in the work. To facilitate the production of a cutter of the right diameter and also to make it possible to lap and oilstone the four-sided cutter so that the cutting edges shall be equidistant from the center, the fixture in Fig. 155 was constructed.

This fixture has an arm composed of two round bars, along which the microscope may be adjusted. The arm is pivoted at one end; at the other end are a pair of stops between which the arm with the microscope is free to swing. A scale is placed on one of the bars which carry the microscope, corresponding to a scale on the engraving machine.

HOW THE FIXTURE IS USED

The microscope is slid to the outer end of the swinging arm, and the stops which limit its oscillation are adjusted so that the cross hairs are sighted first on one side of the tracer; upon swinging the arm to the other side, they are then on the other side of the tracer, as seen in Fig. 156, which illustrates the scheme diagrammatically. The tracer itself is placed in a V-block on the base of the fixture, as shown in the photograph. Now to determine the width of cutter to use with this particular tracer, knowing the reduction required, the microscope is adjusted along the scale until the graduation indicating the required reduction is reached.

In the diagram, Fig. 156, the microscope is shown at the left-hand end as first focused on the opposite sides of the tracer, the position of the microscope when adjusted half-way between the original setting and the pivot is also shown, with the cross wires focused on the cutter edge. This setting is, of course, for testing a cutter one-half the size of the tracer. The cutter is held in a draw-in chuck which is placed bodily in the engraving machine, and which is removed with the cutter when the latter is to be ground. This chuck is placed in an indexing fixture, and the microscope is sighted first on one corner of the cutter and then on the one diagonally opposite to see if the diameter is correct. Each cutting edge is inspected by turning around the holder which carries the cutter, one cutting edge after the other thus being brought under the microscope. In this test each face of the cutter is oil-stoned freehand until brought exactly to coincide with the cross hair of the microscope, which shows that both cutting edges are equally distant from the center and must therefore cut alike. The indexing holder, shown in place in the fixture in Fig. 155, is not the device referred to, but is of somewhat the same nature.

CONVENIENCE OF THE FIXTURE

It will be evident that this fixture is very handy in making comparative tests. No matter what the curvature of a tool may be, the adjustable

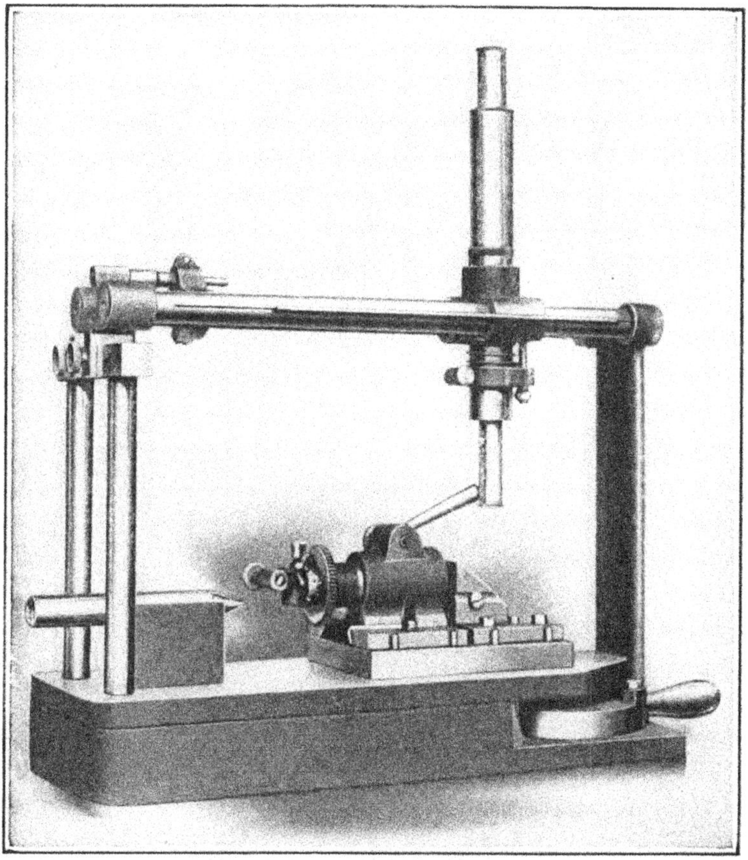

FIG. 155. — A SPECIAL MICROSCOPE FIXTURE

stops may be set so that when the microscope arm is swung to either side one of the cross hairs will come tangent with the curve; any tool which it is desired to duplicate may be satisfactorily tested in this manner. The settings that are possible along the scale also make it just as con-

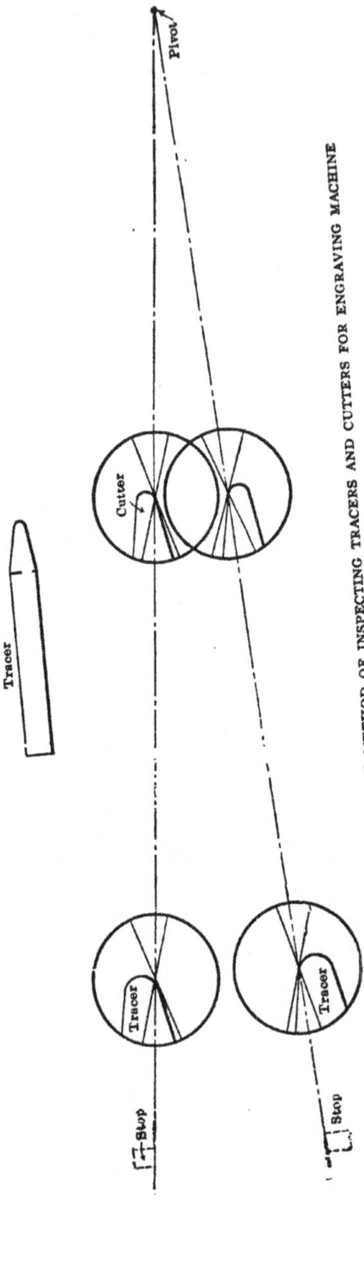

FIG. 156. — DIAGRAM SHOWING METHOD OF INSPECTING TRACERS AND CUTTERS FOR ENGRAVING MACHINE

venient to produce a tool smaller or larger by any required amount than the model. A micrometer head and spindle are attached to the outer end of the swinging bar, so that the microscope may be adjusted by thousandths along the bar on which it is mounted.

This is also a convenient fixture for testing taps and other threaded work, graduated scales, etc., as the microscope may be shifted by means of the micrometer screw and give settings varying by thousandths and fractions of thousandths.

A HANDY MICROSCOPE

The instrument illustrated in Fig. 143 is a general-utility microscope which should find a place in the toolmaking and tool-operating field. Aside from the several ways already mentioned in which it may be used, there are any number of places and machines where it should be a convenient instrument. In testing the running accuracy of a milling cutter, for instance, it may be sighted on the corner and top of the tooth and the spindle then pulled around by hand, while each tooth is examined successively to see if it be high or low. Another place where this instrument could be used to advantage is in setting a planer tool to hight, first sighting on a scale on the desired graduation and then on the point of the tool. Owing to the V in the base block, it is possible to place the instrument on a boring bar and swing it about in an arc of a circle while inspecting a box or cylinder which is to be bored by the bar.

SETTING A SURFACE GAGE

While the job shown in Fig. 141 is the locating of an existing line parallel to an existing surface, the microscope may be used for setting a surface-gage scriber to a given hight by a standard scale (placed vertically), so as to produce a line parallel to an existing surface. Fig. 157

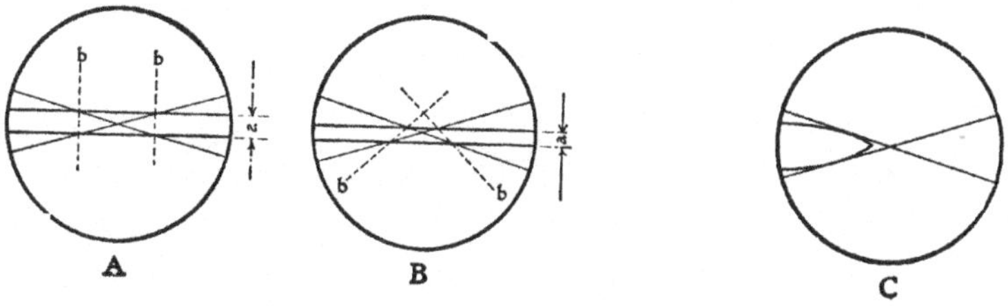

FIG. 157. — SETTING A SURFACE GAGE BY THE MICROSCOPE

shows the idea and also brings out the advantage of having the wires crossed at such a sharp angle. Suppose the lines on the scale when sighted in the microscope have an apparent width *a* in the sketches at *A* and *B*, Fig. 157. If the intersection of the cross hairs is at the middle of the width of the line on the scale, the hairs will cross the upper and lower edges of the line at equal distances from the center, as in the sketch at *A;* if we conceive lines like *b b* passing through the points of intersection of the cross hairs with the edges of the scale line, these lines will be vertical, as indicated. The sketch at *B* indicates the effect when the cross hairs are not central with the edges of the scale line, the lines *b b* here being at a considerable angle with the vertical.

When the microscope is adjusted central with the desired line, the scriber block may be adjusted until the point of the scriber lies between the cross hairs of the microscope, as at *C*, in which position it will be at the desired hight for scribing the line on the work. It will be apparent that the scriber must be in a horizontal position when setting the point properly by the opening between the cross hairs.

CHAPTER XIV

The Microscope in the Manufacturing Plant

In Chapter XIII the several illustrations presented show a number of applications of the compound microscope to the work of the toolmaker. It is the purpose under the present heading to illustrate some further advantages of the microscope in connection with manufacturing processes. The instruments and processes shown should also be of interest to toolmakers as suggesting numerous other applications of the microscope to fine work.

The Lanston Monotype Machine Company, of Philadelphia, Penn., has for several years made extensive use of the compound microscope for facilitating the production and inspection of certain very accurate members entering into the construction of their casting machine. While there are in this machine many other parts the workmanship on which is of the finest character, the members particularly referred to here are the matrices from which the type is cast; these parts and the hardened-steel punches for sinking the characters in the matrix bodies pass in their production through some very accurate and interesting operations.

A typical microscope and its stand, as used in this establishment, are shown in Fig. 158, mounted on a bench; certain details of construction are illustrated by the line drawing, Fig. 159. In the photographic view a matrix will be noticed in position for inspection under the microscope tube.

THE MONOTYPE MATRIX

The matrix is sketched in Fig. 160 about double size, the body being $\frac{2}{10}$ inch square for type sizes from 4- to 14-point, and $\frac{13}{32}$ inch high. The punch is sunk into the matrix (which is of hard copper) 0.065 inch in making the impression therein for the character afterward to be formed in reverse on the type. The matrices (225 in number) are carried in a matrix case in the casting machines. There are 15 rows of 15 matrices each in the case, and each row is carried upon a wire passing through holes drilled crosswise of the matrix bodies. The case may be moved horizontally in two directions at right angles, to bring any matrix into casting position over the mold. Each matrix is centered with the mold by a taper pin fitting the hole in the end of the matrix, the movement of the case being automatically controlled by a perforated paper ribbon which has been previously punched on a keyboard machine. Now in order that each character on the type when cast shall be of proper size, with each line composing the letter or other character of the correct width, and each character correctly located on the type body — not only squarely positioned thereon, but also at the right distance from the edges of the body, so that perfect alinement will be assured between all the characters — it is absolutely essential that the punches (which are really

FIG. 158. — MICROSCOPE WITH WORK IN PLACE

master type of hardened steel) shall be made with the greatest accuracy. The characters must be dead square in their position on the punch end, each and every line forming the character must be of the exact width desired, and the lines forming the top and bottom and the sides of the character must be at a given distance from the sides of the punch body.

INSPECTING THE MATRIX

After the punch has been finished in an engraving machine and hardened, it is ready for sinking the impression in the copper matrix, which operation is performed in a power press. After the impression is formed

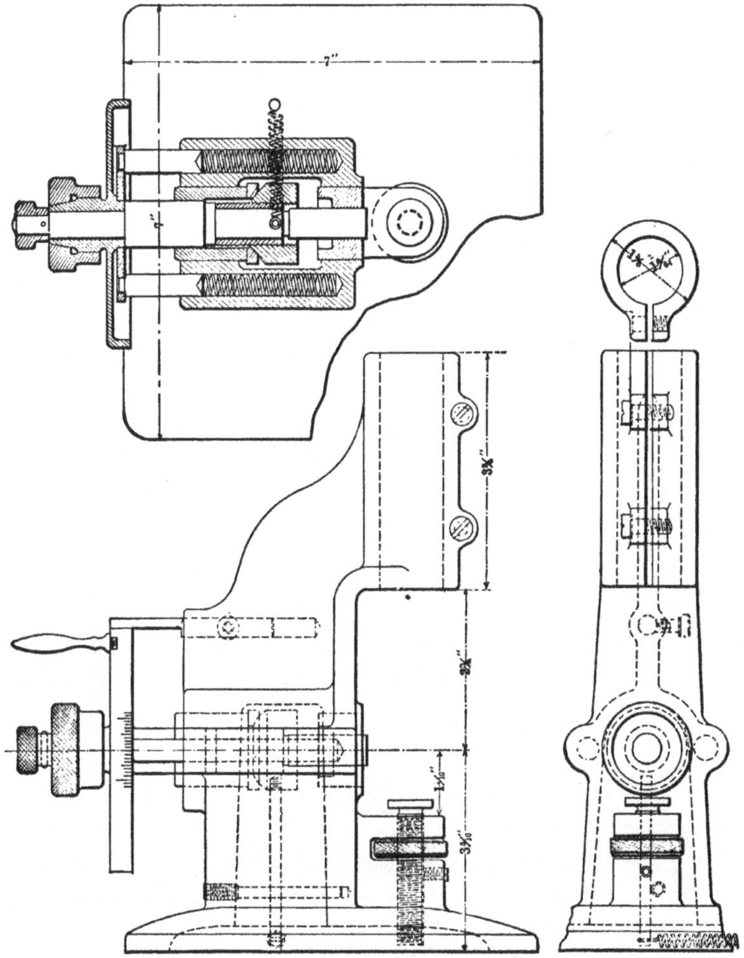

FIG. 159. — CONSTRUCTION OF MICROSCOPE STAND

the conical centering hole is finished in the end, the hole for receiving the matrix case wire drilled through the matrix body and the copper matrix milled off to exact length and to leave the exact depth of impression required, in an ingenious special machine. The matrix is then ready for inspection under the microscope.

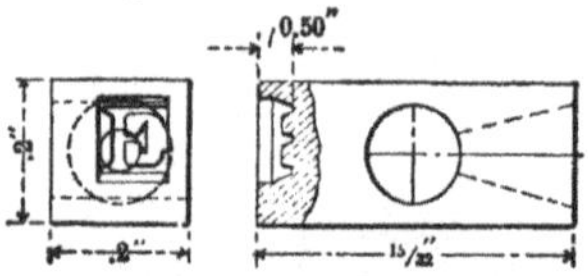

The Matrix

FIG. 160. — MONOTYPE MATRIX

The cross hairs in the microscope are at 90 degrees to each other, as seen at *A*, Fig. 161. The test for the squareness of the base line of the character with the matrix side is indicated at *B*. *C* shows the letter as seen under the microscope, with the crossbar, say of a letter *H*, being tested against the cross hair of the instrument. The matrix rests, as shown in Fig. 158, on a lapped plug in the base of the instrument; one side is carried against the lapped end of a horizontal rest adjusted in and out by the micrometer screw and dial at the rear of the stand. The dial is graduated to read to ten-thousandths of an inch, and it is obvious that very minute measurements may be readily made by the combination of microscope and micrometer.

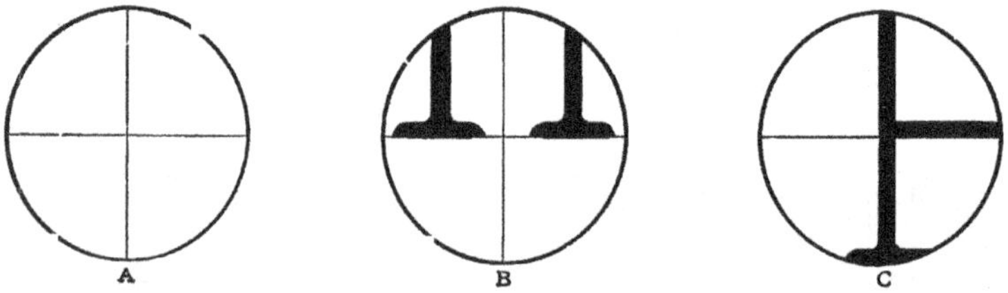

FIG. 161. — USE OF THE MICROSCOPE CROSS HAIRS

The method of measuring the width of a line in the character is also indicated at *C*, Fig. 161, where one of the cross hairs of the instrument is shown coincident with the edge of the cross line in the letter. By turning the micrometer screw until the opposite edge of the line being measured coincides with the microscope cross hair, the width may be read at once from the micrometer dial.

The knife-edge square shown in front of the microscope stand in Fig. 158 is used in testing the squareness of the sides of the work with one another. This square is used quite generally in the Monotype shops.

The microscope stand construction will be sufficiently clear from the photograph and the line drawing, Fig. 159, without detailed explanation. Ordinarily, while in use, the base of the stand is covered with a piece of felt, so that if a matrix drops accidentally, there will be no risk of its being injured.

MICROSCOPE WITH TWO SCREWS

Fig. 162 illustrates another microscope for matrix inspection, which has two micrometer screws at right angles to each other.

FIG. 162. — MICROSCOPE WITH TWO MICROMETER SCREWS

The photograph shows plainly the method of locating the matrix under this microscope, where it rests in a notch cut in a right-angle shoe, which in turn rests against the ends of the two micrometer screws. This feature is also shown in the right-hand sketch in Fig. 163.

It is obvious that with this microscope and its two screws it is an easy matter to measure the matrix impression from one side to the other, or from the top to the bottom of the character, without changing the position of the work in the V-block in which it is held squarely. It is also a convenient instrument for testing the correctness of the position of the different lines of the impression relatively to the edges of the matrix body.

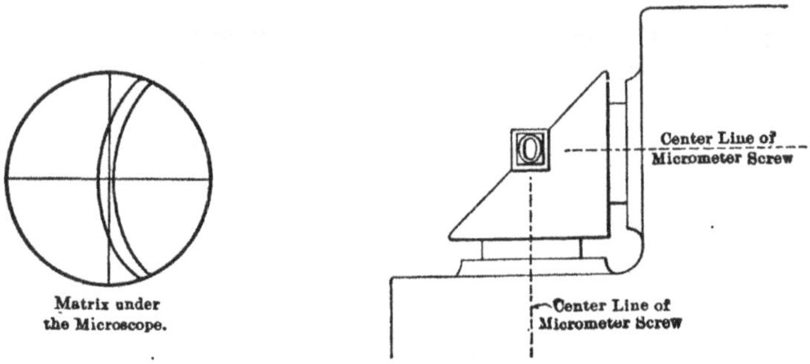

FIG. 163. — PLAN OF MICROSCOPE WITH MATRIX IN POSITION

To illuminate the impression in the work sufficiently to allow easy manipulation under the microscope, a small magic lantern is employed, into which is dropped a regular incandescent light bulb, the reflector then lighting the work nicely. A number of illuminating devices thus arranged will be noticed in other photographs in this chapter. In this particular case the lantern was moved prior to the taking of the picture.

OTHER MICROSCOPES IN THE MATRIX DEPARTMENT

Fig. 164 shows in a striking manner the extent to which microscopes are used in connection with the production of the matrices, there being on the bench in this view at least a dozen such instruments. With the microscopes in this department the inspection of the impressions in the matrix is carried on with facility, as the method of lighting, which is clearly indicated and which has already been referred to, enables the operator to view the bottom of the impression plainly through the microscope, as he examines the various portions of the character stamped in the matrix. As all these tests are carried along from the bottom of the impression, it is obvious that the type cast therein must be perfect on the face.

It should be stated that the row of machines shown to the right, in Fig. 164, is a part of the group of presses used in forming the matrix impression.

ANOTHER INTERESTING OPERATION

Another application of the microscope is represented in Fig. 165, which shows a special arrangement for the inspection of display matrices.

FIG. 164. — ROW OF MICROSCOPES IN MATRIX DEPARTMENT

The blanks in which the matrix is formed are about $1\frac{5}{32}$ inches long by $\frac{25}{32}$ inch wide. The character is formed in this matrix by a special process, the details of which need not be taken up at this time.

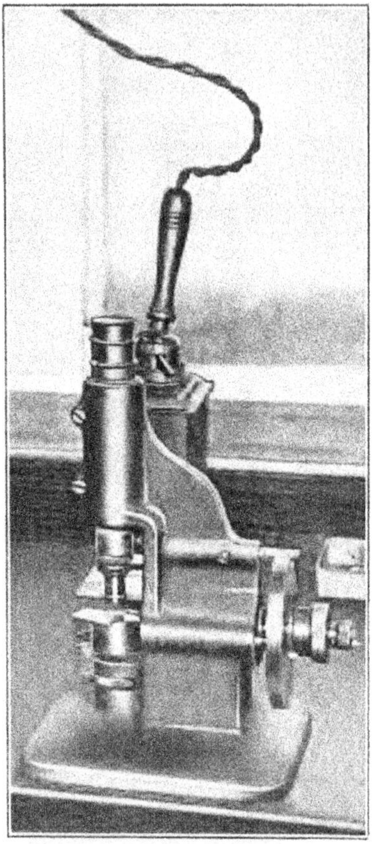

FIG. 165. — MICROSCOPE FOR INSPECTING DISPLAY MATRIX

With the display matrix, as with the other form of matrix already shown, it is important that the impression from which the type is to be cast should be square with the edges of the blank and located at a certain distance from the ends and sides. The work is placed under the microscope, Fig. 165, for inspection. The micrometer dial on this instrument like those on the other microscopes reads to 0.0001 inch, and forms a very convenient means of measuring the work accurately.

CHAPTER XV

Making a Set of Accurate Index Dials

The problem of originating an accurate index dial or plate is one of the most interesting and difficult of the many special undertakings within the province of the toolmaker. Usually the origination of a single dial only is involved; but where there are two or more to be made, each dial is generally handled practically as an independent proposition. If a built-up construction is adopted, the indexing blocks or other members attached to any one of the dials are finished as permanent parts of that particular dial without attempting to treat them as interchangeable members which, if desired, could be applied to any dial in the series without affecting the close degree of accuracy originally provided for.

It is the object of the present article to describe the methods and appliances utilized in the originating of a set of seven index dials, each of which is provided with twelve hardened and ground index blocks so made as to permit any one of the blocks on any given dial to be interchanged with any of its eleven companion blocks or to be placed in any of the twelve positions or stations on any other dial without this change in any way affecting the accuracy of the dial in question. While there is little likelihood of the index blocks ever being changed while in service, the fact that the method of construction makes it possible to drop any block correctly into any station on any dial constitutes a highly interesting feature of the undertaking, although this feature of interchangeability may of itself be said to have added nothing to the cost of doing the work, as it was, so to speak, one of the natural results incidental to the course of procedure followed in the prosecution of the work.

APPLICATION OF THE DIALS

These seven dials were constructed at the experimental shop in New York City of the Remington Typewriter Company. This shop is superintended by George F. Ballou, who is well known to many readers in connection with his work in constructing the Rogers-Ballou dividing engine and other precision undertakings of the finest character. Each dial forms a part of an automatic machine used in performing a series of delicate operations upon a U-shaped typewriter part, measuring about $1\frac{1}{8}$ inches in length. The index dial, which is mounted on a horizontal spindle, carries on the side opposite the index blocks twelve vise blocks for carrying the work, and at each advance of the dial four spindles located at either side advance the tools which operate upon two sides of four pieces simultaneously. The dial is rotated by a twelve-tooth ratchet on the spindle and is of course held fast by the locking bolt during each operation on the work. The machine once assembled provides for no adjustment either as to center distances between spindles or re-timing of index and spindle movements. The mechanism is constructed to take care of the accurate machining of this particular piece and no loophole is left for improper readjustment of any important member. The operations on the work are of a peculiarly sensitive nature, and it is essential that the successive index movements be very exact in order to assure the proper action of the tools and the necessary degree of precision in the product.

Six of the set of seven dials are shown in Fig. 166; the seventh was in operation in its machine at the time the photograph was taken, hence could not be included in the group. The dial just to the left of the center is represented from the vise-block side. These blocks had been removed from the dials for some purpose and had not all been replaced when the exposure was made. However, the three which are shown give an excellent idea of the method of attaching to the dial; and the larger view in Fig. 167, which shows opposite sides of two of the dials, brings out clearly the toggle arrangement with which each vise block is equipped for gripping the piece of work. The latter, by the way, is fed into the jaws from a table past which the periphery of the dial turns; at the first indexing movement after the admission of the piece a locating device sets the work firmly in place in the holder, the second indexing movement causing the jaws to clamp it fast and hold it securely until after it has passed the series of tools, when the toggle mechanism is again actuated to open the jaws and release the work, which then drops into a receptacle beneath the dial.

THE INDEX BLOCKS

The manner in which the twelve hardened and ground index blocks are fitted to the dial is well illustrated in Fig. 167. As will be noticed, they rest on an annular shoulder or ring formed on the face of the dial, and each is secured by two dowel pins and two fillister-head screws. The index notches are of the well-known form, having one radial face and an opposite face at an angle of 30 degrees; the notches are formed between the unbroken radial end of one block and the 30-degree shoulder provided on the block adjacent.

The disk on which the blocks are carried is of cast iron, 16¾ inches in diameter and finished all over, with a hub about 5 inches long bored to 2½ inches and keyed on a hardened and ground steel arbor on the opposite

FIG. 166. — A GROUP OF ACCURATE INDEX DIALS

FIG. 167. — FRONT AND REAR VIEWS OF A PAIR OF INDEX DIALS

end of which is secured the ratchet for rotating the dial. A detail of the blank disk finished and ready for the blocks is shown in Fig. 168. A block detail is represented in Fig. 169.

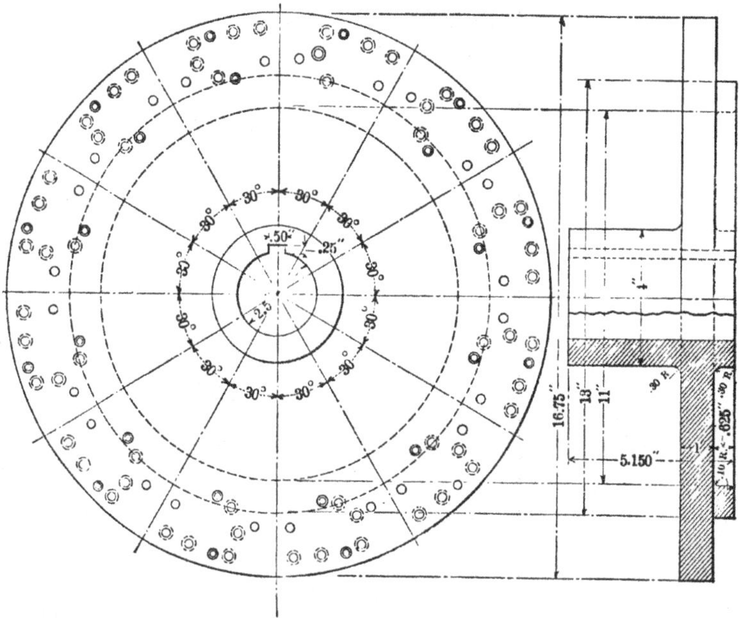

FIG. 168. — THE BLANK DISK

ROUGHING OUT THE INDEX BLOCKS

These blocks are approximately $4\frac{1}{4}$ inches long over all, and $\frac{5}{8}$ of an inch thick; the material is Jessop steel. They are secured to the disk by two $\frac{1}{4}$-inch straight dowel pins fitting holes *A A*, and two $\frac{3}{8}$-inch fillister-head screws fitting soft-steel bushings at *B B*, which are pressed firmly and permanently into place. The object of these bushings is to provide a means of accurately locating and sizing the holes for the screw bodies and heads without the necessity of grinding these holes out after the hardening of the blocks. Therefore in drilling the blocks, holes for the bushings were drilled, reamed and counterbored, and after the blocks were hardened the unhardened bushings were pressed into place, to be again drilled, reamed and counterbored for the screw bodies and heads in a later operation.

The 84 blocks necessary for the seven dials were cut roughly to length, machined carefully on the sides to within a few hundredths of the required thickness, then placed nine at a time on the lathe face plate shown to the left in Fig. 170. Here, while held by straps, a few of which are represented in position, the outer and inner edges were turned to roughing dimensions, preparing the blocks for the drilling operations to be performed in the jig. This jig will be seen at the center of the group of tools in Fig. 171.

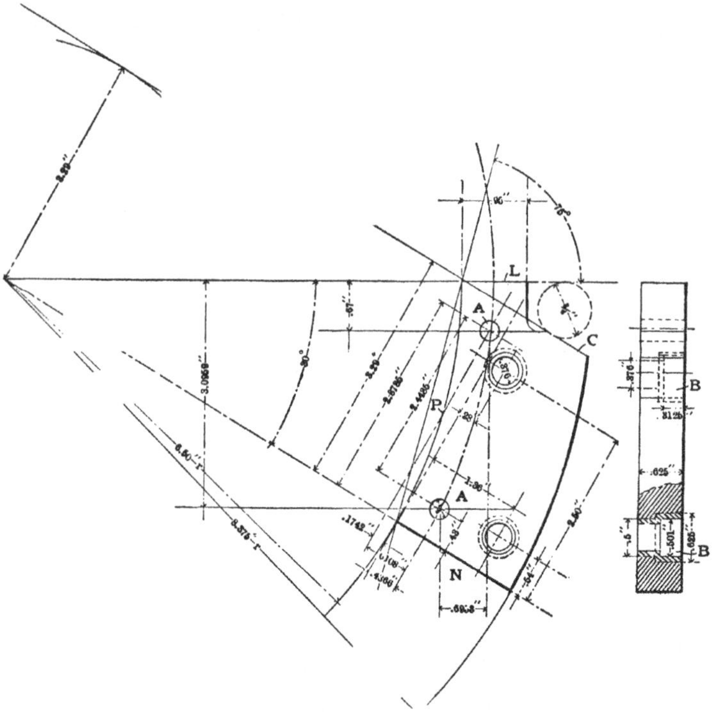

FIG. 169. — DETAIL OF INDEX BLOCK

THE DRILLING AND REAMING JIG

It should be stated here that the index blocks were passed twice through this drill jig. For the first operation in the jig the two pins shown at one side of the center were removed, the bushings in which they are shown in the half-tone being utilized for drilling the two straight dowel-pin holes *A A*, Fig. 169. The work during this first operation in the jig was located against hardened-steel plates at the side and end. The dowel-pin holes were drilled and reamed a few thousandths under size and the two holes at *B B* for the bushings to receive the fillister-head

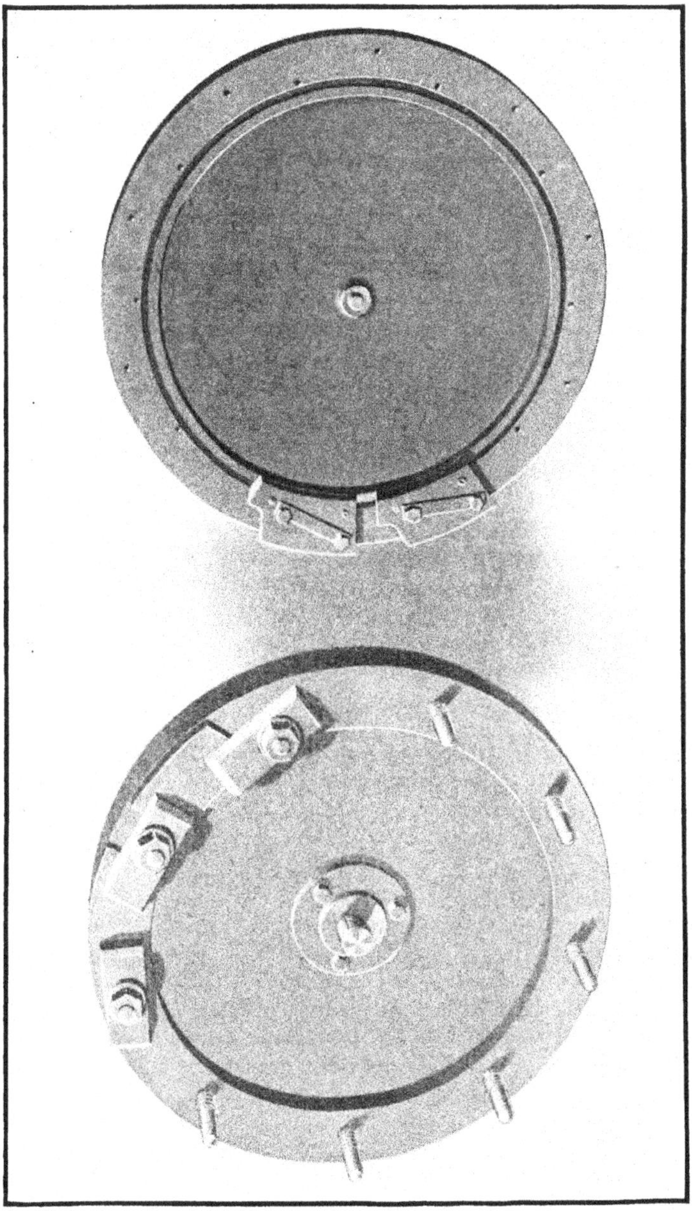

FIG. 170. — FACE PLATES FOR TURNING AND GRINDING THE INNER EDGES OF THE INDEX BLOCKS

screws were drilled, reamed and counterbored nearly $\frac{1}{8}$ inch larger than the actual size of the screw heads and bodies. At the same setting a hole was drilled through at one end of the block to form a clearance space to facilitate the finishing of the 30-degree bevel surface at *C*. After this end had been shaped out and the blocks had been hardened, two soft-steel bushings turned on an arbor slightly smaller than the $\frac{3}{8}$ screw bodies were pressed into the two enlarged holes referred to above; the blocks were then ready for grinding on the faces, this being accomplished on the magnetic chuck in the surface grinder, where the faces were finished parallel and the desired thickness of $\frac{5}{8}$ inch obtained. The second operation performed in the drill jig consisted in reboring and counterboring the soft-steel bushings to suit the fillister-head screws; it will be referred to again at the proper point in this description. Before leaving the jig for the moment, however, attention should be drawn to the fact that it is so bushed from opposite sides as to permit drilling of the work from one side and reaming from the other.

After the index blocks had been handled in the jig for the first time, the two $\frac{1}{4}$-inch holes *A A* in each were utilized as locating and reference points to facilitate the accurate finishing of various surfaces. It was therefore essential that these two holes in each of the 84 blocks should be as accurately spaced and as straight and true to size as it was possible to make them. This necessarily involved some exceedingly nice grinding and lapping operations in the case of each pair of holes, and some details of the procedure should be of interest here.

THE MASTER TEMPLETS

First a soft master templet was made, which was used in making a hardened master, which was in turn utilized for locating the work properly on the bench lathe for the accurate finishing of the two $\frac{1}{4}$-inch holes. As explained in other sections of this book, there are several methods known to the toolmaker for spacing holes accurately in a jig or a templet, some involving the use of buttons and indicators on lathe or miller, others depending on entirely different means for attaining the required degree of accuracy. One method, which is simple though highly satisfactory, makes use of the dial readings on the screw of the milling-machine table, yet does not depend at all upon the accuracy of the screw for the nicety of the results obtained.

Thus in boring say two $\frac{1}{2}$-inch holes 2 inches apart in a plate, this method would involve first the securing of the work to the table, and the boring of one hole to a size somewhat below $\frac{1}{2}$ inch. Before starting this hole it is assumed that the back lash is out of the screw, the table and other members snugly adjusted and the graduated dial set at zero. Now the table is fed along to a distance of 2 inches, as recorded by the dial on

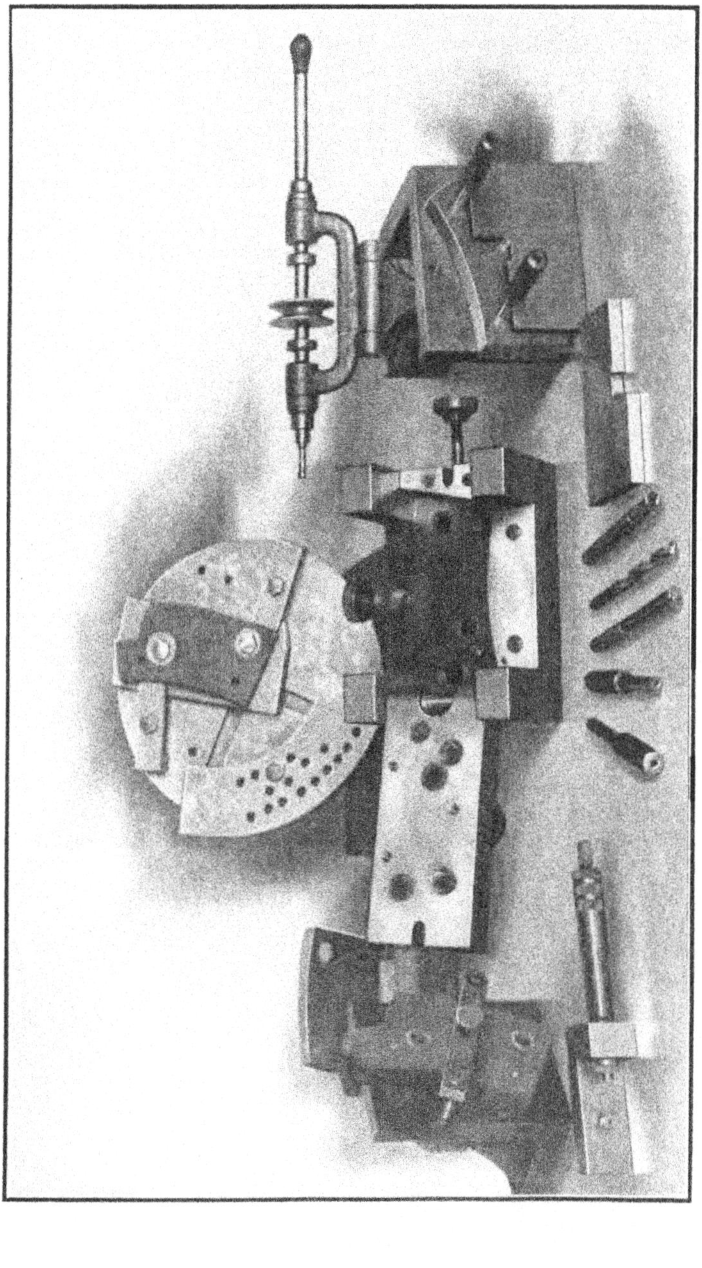

FIG. 171. — A GROUP OF INDEX-BLOCK TOOLS

the screw, and the second hole bored to the diameter of the first. Two plugs are inserted in the holes, and a careful measurement made over them with micrometer or vernier to determine the actual variation from the required center distance of 2 inches, the diameter of one plug, of course, being subtracted from the reading of the instrument to give the reading for center distance. Say the holes are a thousandth or so too near or too far apart: The table is readjusted in the required direction by the desired amount, and the hole rebored to the right size, when, if desired, another plug may be put in for a second test measurement. If the center distance is now correct, the table is moved back by the screw to bring the first hole into position for reboring to size. In running the table back, the screw is turned some little distance past the zero mark to which the original setting was made, and then turned ahead again until the dial is once more at zero. This takes out the back lash and the table is again in the position it occupied at the time the first hole was bored undersize. This hole is now rebored to $\frac{1}{2}$ inch, and if the customary care has been used in the manipulation of the machine and boring tool, the two holes will be found to be correctly located relatively to each other.

This method as outlined was the one adopted in boring the two important holes in the soft templet for the index blocks, these holes being made $\frac{7}{16}$-inch diameter and to the exact distance apart as computed from the drawing, Fig. 169. From this templet, which was about $\frac{3}{8}$ inch thick and $2\frac{1}{2}\times4\frac{1}{2}$ inches over all, a block of the same size, but about $\frac{5}{8}$ inch thick, was laid out for two $\frac{3}{8}$-inch holes, which were drilled a little undersize. Two other holes were drilled, corresponding in position to two tapped holes in the thinner templet, these latter holes being adapted for a couple of screws to be used in fastening the two templets together. Also two holes were tapped in the thicker templet to allow the index blocks to be attached later on.

The $\frac{5}{8}$-inch templet was afterwards hardened and ground parallel on its faces, and was then ready to have the two holes corresponding in position to the dowel-pin holes in the index blocks ground out to correct size or $\frac{3}{8}$ inch, and to correct distance apart. This operation was performed in a bench lathe by the aid of a diamond-charged grinding disk.

CORRECTING THE HARDENED MASTER

The two templets were put together, the hard plate on the soft one as indicated at D and E, Fig. 172, and the two screws put in to hold them securely, the holes to be ground out in the hard templet E being located as nearly central as possible over the holes in the soft master D. The face plate F, of the lathe, was first carefully trued up in place on the spindle and scraped flat to a surface plate; the hole bored in its center was fitted with a steel plug G which, after being put in place, was ground at the outer

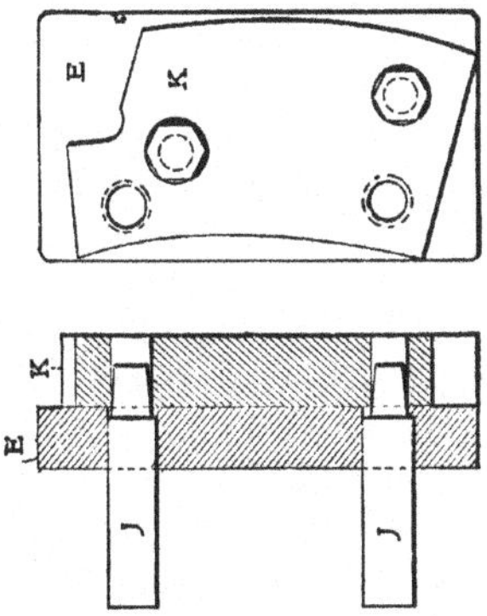

FIG. 173

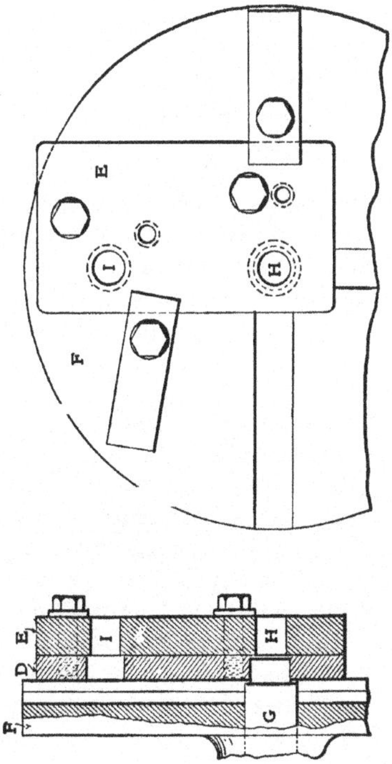

FIG. 172

LOCATING THE WORK ON THE BENCH-LATHE FACE PLATE

end to fit snugly in the holes bored in the soft master templet *D*. The two templets were then put on the plate with the master *D* slipped over the locating plug, a couple of straps holding the two templets in place. A counterbalance was put on the face plate, and in order that there should be no danger of the weight springing the plate when clamped in place its under surface was carefully scraped to a good bearing prior to mounting on the plate. The general arrangement of the work and face plate was then as indicated in the sketch in Fig. 172. The grinding attachment used on the job is seen to the right of the face plate in the group in Fig. 171.

After hole *H* had been ground out with the diamond-charged wheel to within about 0.0001 inch of the required size, both templets were removed from the face plate, and turned end for end to bring hole *I* into position for grinding. It is evident that this method of locating hardened templet *E* by means of the accurately made soft master *D* assured the correct center distance being obtained between the two holes in *E*.

After the two holes had in this way been brought to within about 0.0001 inch of correct size, the hardened master was placed in a sensitive drill press, the table of which was carefully squared up; with a copper lap mounted on a spindle running true in the chuck the holes were lapped out dead to size, the master now being ready for its work of locating the index blocks on the face plate so that their dowel-pin holes could be ground out to ¼ inch.

GRINDING OUT THE HOLES IN THE INDEX BLOCKS

In locating an index block on the hardened master templet, two plugs with bodies fitting the holes in the latter and with the ends very slightly tapered were pushed in from the back of the master as shown in Fig. 173, the taper ends of these plugs *J* entering the index block *K* far enough to center it, even though the original distance between the dowel-pin holes had changed somewhat in the hardening operation. Two screws were then put in to secure the index block to the master templet *E*, and after the locating plug at the center of the bench-lathe face plate had been reground to the size of the two locating holes ground out in *E*, the work was placed on the face plate. This plate, with the hardened master templet and an index segment clamped in position, is shown just behind the jig in Fig. 171. The two ¼-inch holes in the index block were ground out in the bench lathe in the same way as the holes in the master plate were treated, and a very slight amount was later taken out in the lapping under the drill spindle to bring the holes dead straight and to the size of the standard plug. The 84 blocks in the series were put through this process one at a time, and some idea of the care taken in the work may be obtained from the fact that no less than two months' time was required in the finishing of the 168 holes.

FIXTURES USED IN GRINDING THE ENDS OF THE BLOCKS

To hold the index blocks in the proper position for grinding the ends to the radial lines at an angle of 30 degrees with each other and the proper distance from the dowel-pin holes, two small angle plates shown in the half-tone Fig. 171 were made. These were finished with great care to bring the surfaces and edges perfectly square with one another. After the angle plates were made, came the problem of locating in each, two pins upon which to locate the work from the ¼ holes ground out in the manner described.

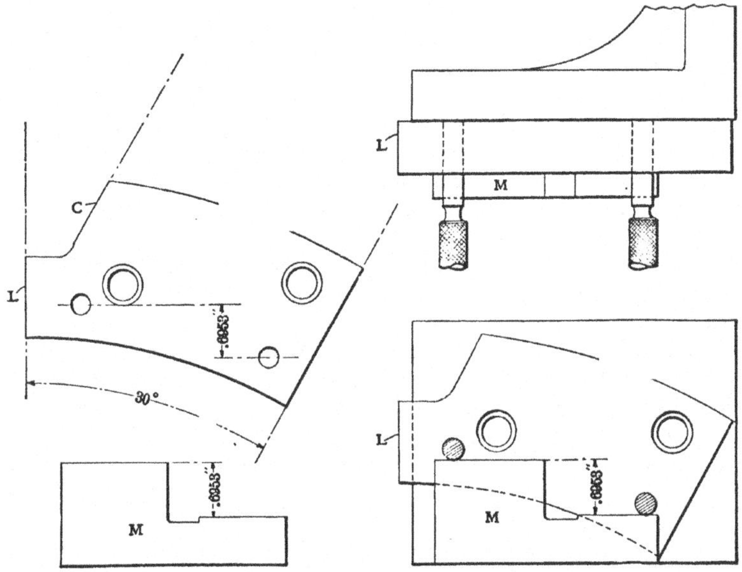

FIG. 174. — LOCATING INDEX BLOCK ON ANGLE PLATE FOR GRINDING SHORT END

The angle at the left of the group in the engraving is shown with an index block in position for grinding the shorter radial end. The correct location of the pins in this case was obtained as follows:

Reference to the detail of the index block in Fig. 169 will show that the various dimensions given include a measurement — 0.6953 inch — between the center lines of the two dowel-pin holes *A* drawn perpendicular to the edge *L* to be ground. A templet like *M*, Fig. 174, was made having an offset corresponding to this dimension, this templet being ground and lapped on the upper and lower edges to bring these parallel and give the exact drop required. The angle plate was placed on its edge with the templet *M* in front, two plugs fitting the ¼-inch holes in an index block were put into place and the block itself placed against the face of the angle plate with the two plugs resting on the edges of the templet as shown in the sketch. The index block was then carefully clamped to the angle plate and the plugs removed. The job was taken to the sensitive drill; a spot drill whose body was a nice fit in the holes in the index block and whose cutting end was flatted and point thinned down to give a neat cutting edge was passed through the index holes in the block to spot the surface of the cast-iron angle plate. A slightly smaller twist drill was then run completely through the angle plate, leaving a small amount of metal to remove by reaming. The reamer which finished the holes was made a close fit in the index-block holes and adapted to cut on the end only. This reamer and the spotting drill will be seen in the foreground of Fig. 171. When the index block which thus served as a jig was removed, two hardened, ground and lapped pins were put into the angle plate, and with the addition of a swinging clamp for securing the work the angle-plate fixture was ready for use in the surface grinder. In use it was of course placed upright, as indicated in the half-tone Fig. 171, bringing the surface *L* of the work to be ground to the top, and parallel to the base of the angle plate.

In arranging the other angle plate or fixture for holding the index block for the grinding of the longer radial surface *N*, Fig. 169, a similar method was followed, another templet *O*, Fig. 175, being required in which the amount of the drop at the edge equaled 0.6108 inch, the distance between the dowel-pin holes drawn perpendicularly to the longer edge *N*. This angle plate is shown in Fig. 175, also in Fig. 171, with an index block in front supported on its templet ready for clamping preliminary to the spot drilling, drilling and reaming of the holes for the ¼-inch locating pins.

APPLICATION OF THE FIXTURES

These angle plates used as grinding fixtures formed an excellent means of testing the accuracy of the locating holes in the index blocks, for the slightest deviation from the center distance with the holes lapped dead to size would have prevented the blocks in question from going into place on the fixtures. The accuracy obtained in the spacing of the work is indicated by the fact that every one of the 84 blocks fitted properly over the pins, and for that matter would go either side out over the pins on either angle plate.

On these two fixtures in a Brown & Sharpe surface grinder the ends were brought to the dimensions given in the detail Fig. 169, from the dowel-pin holes. Measurements were obtained by means of the special micrometer gage shown in the foreground of Fig. 171, the pin in this gage fitting the ¼-inch holes in the work and allowing the end of the spindle to come down squarely on the end of the piece. A vernier plate was attached to the bracket behind the hand wheel for adjusting the wheel slide of the grinder so that the graduations on the hand wheel reading to thousandths could by the vernier be read to ten-thousandths. By means of this attachment very fine adjustments of the cut were possible, so that when the work left the grinder it was correct so far as the ends were concerned. The 30-degree surface *C*, Fig. 169, on the index block, corresponding to the sloping side of the lock bolt, being parallel to the end *N*, was ground by resting the block on the latter end and clamping against an angle plate.

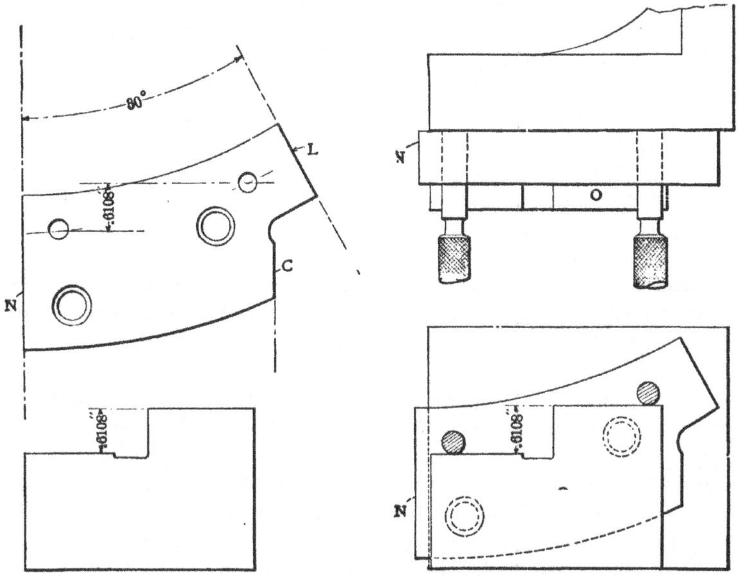

FIG. 175. — LOCATING INDEX BLOCK ON ANGLE PLATE FOR GRINDING LONG END

GRINDING THE INNER CURVE OF THE BLOCKS

Now we come to the grinding of the index blocks along the inside curve *P*. For this operation the face plate shown to the right in Fig. 170 was made. This plate was fitted to a lathe, carefully trued up in place, and the plug fitted at its center ground true to an even dimension. The cutting of the annular slot for locating the work constituted the most difficult part of the undertaking. The center line of this slot is on a radius of 6.95 inches in accordance with the figures on the detail sketch of the index block. The width of the slot being of no importance of itself, every attention was paid to getting its actual center the exact distance from the center of the plate. After the slot had been cut with the tool set as closely as possible at the desired point, the distance from both inside and outside edges of the slot to the outside of the central plug was measured, and the actual deviation from the desired radius was easily learned by a simple calculation. The tool was then adjusted to dress out one side of the slot according to whether the radius was long or short, and measurements were again taken. After a few trials a perfectly concentric slot was produced whose center line was exactly to the required radius, and whose sides were exactly parallel.

A series of pins was next made with bodies fitting the ½-inch holes in the index blocks, and with heads fitting the annular slot turned in the face plate. Ten blocks were then mounted on the plate at a time in the manner indicated in Fig. 170, a series of holes being drilled and tapped in the plate for the reception of the screws for holding the work in place. A grinding attachment rigged up on the lathe completed the outfit, and the grinding out of the inside curve of the blocks was a comparatively simple matter, although, naturally, the internal diameter measured across from block to block had to be gaged with considerable care.

SECOND OPERATION IN THE JIG

After leaving this fixture the blocks were returned to the drill jig, where the holes in the soft-steel bushings were brought to size, and counterbored to the size called for in the drawing. In this second operation in the jig the work was located over the two hardened pins shown in the view in Fig. 171, and this again was a good test of the accuracy of the workmanship on the pieces as well as on the jig itself. The bushings originally used in the drilling and reaming of the ½-inch holes were changed for a pair suitable for the 0.376-inch holes, to which the soft bushings for the fillister-head screws were finished in this operation; drilling and reaming being accomplished from opposite sides of the jig. Upon counterboring the soft bushings to 0.501 inch, the blocks were ready for assembling on their dials.

ASSEMBLING THE INDEX DIALS

This operation was accomplished with the aid of the cylindrical skeleton fixture shown to the left of the drill press in Fig. 176. The blank dial was placed in this frame right side up and eleven blocks carefully cleaned of all dirt and lint placed approximately in position on its face, drawn up to one another and lightly pressed by their set-screws in the periphery of the fixture. The twelfth index block was then placed in the gap left in the series, and, in the case of each dial, this last block when slipped in came just in contact with its fellows, and yet allowed a thickness of 0.001 inch tissue paper to be drawn out from between each end of the curved inner edge or arc of contact, and the annular shoulder on the disk to which the blocks were fitted. This means that when the screw was brought against this last block it set the entire series of indexing segments snugly together ready for the drilling of the disk on which they were mounted.

Before starting operations in the drill press the table was squared up and the drilling and reaming of the dowel-pin holes and the drilling and tapping of the holes for the fillister-head screws were proceeded with; no pains being spared to insure the accuracy of the results. The holes were all spotted by the spot drill, and the pin holes were drilled through with a slightly smaller twist drill, and reamed with a reamer fitting accurately in the holes in the index block and cutting on the end only. The drilling and tapping of the other series of holes were done carefully and no opportunity was given either drill or tap to "run" during the process. The dowel pins used were hardened, ground and lapped straight and true and to size, and the screw holes in the blocks provided only 0.001 clearance for the bodies and heads of the screws, so the necessity for careful handling of the work will be obvious.

With the pins and screws in place the job was ready for the final operation of bringing the outside of the index blocks to the required diameter, which operation was readily performed by mounting the dial bodily in the grinding lathe.

This completed the work on the indexing segments. To be fully appreciated they must be seen and tested by the observer, who will not fail to admire the beautifully sharp, clean joints of the index blocks when examined under a glass, and the exquisite fit of the different blocks when removed from their places and tested on the several fixtures which aided in their production. The blocks slide over either pin in either of the surface grinding fixtures absolutely without shake, and yet pass with the same nicety of fit and either side out over the two pins which constitute the locating members on these fixtures. They fit either side up over the pins in the drill jig; they measure alike over every point where accurate measuring instruments can be applied; and there is no index block in the set of 84 but what can be removed from its dial and put into the place of any other block on any dial with the certainty that it will fit at every point.

PUTTING ON THE VISE BLOCKS

In conclusion reference may be made briefly to the fixture shown on the drill press, in Fig. 176. In this view the work is shown with the reverse side of the dial up, as actually handled while the vise blocks for carrying the typewriter parts through the automatic machine were being located in position. The fixture was constructed with a locking bolt fitting the index notches of the dial itself, so that the latter might be locked at the different points, as it was turned to bring one section after another under the drill spindle. In designing the device, provision was made for clamping the vise block to be attached, to a positive locating block fixed in the proper position relative to the locking bolt at the under side; with the vise block thus held fast, the spot-drilling, drilling, reaming and tapping operations were carried along with the same degree of care as was em-

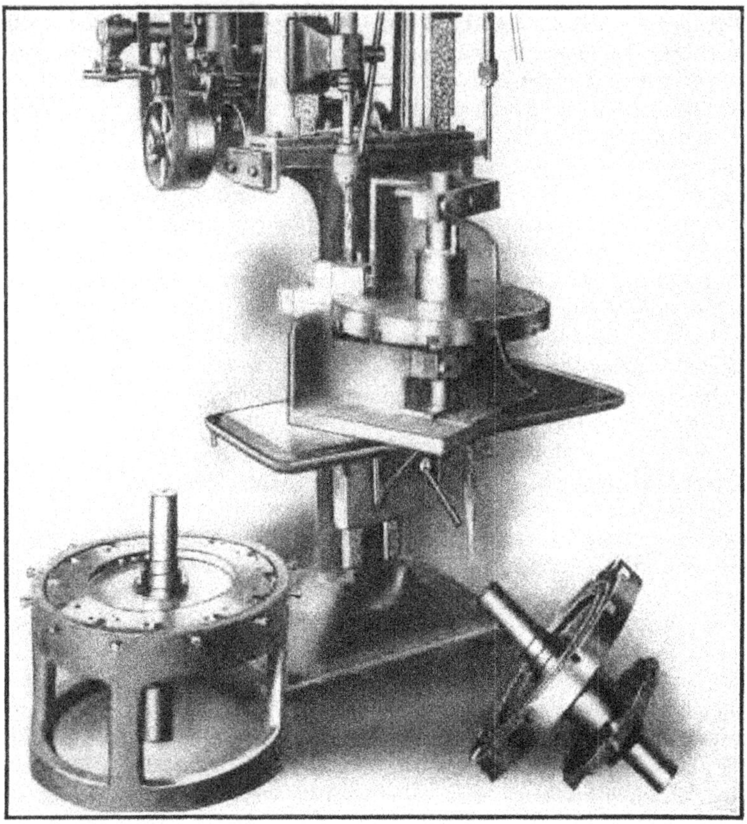

FIG. 176. — THE ASSEMBLING FIXTURES

ployed in the case of the index blocks. As will be noticed, the engraving shows the dial completely drilled. It was completed some time before the photograph was taken, and was merely set up in the fixture with only one block in position to indicate the method of handling during this operation.

CHAPTER XVI

INSPECTING TOOLS WITH THE TEST INDICATOR[1]

IN the accompanying illustrations are shown a number of methods of testing and measuring various classes of tools by the aid of a universal test indicator. This instrument, which is manufactured by J. H. Boulet, is shown in Fig. 177, with the different attachments required for indicating surfaces in various planes, reaching into small holes, testing the rear faces of work held in chucks, and for numerous other purposes. Of the two bars shown directly beneath the tool, the one with pointed end is for locating work by a center-punch mark, while the other is of service in testing the alinement of lathe centers, etc. The graduated plate at the end of the shank reads directly to thousandths of an inch.

TESTING A SURFACE OR BED PLATE

For testing the surface of a plate the indicator is mounted, as in Fig. 178, in an upright position on a surface gage fitted with an extension arm of about one-half the length of the plate to be tested. By placing the base of the surface gage at the center of the plate and reaching to its four corners, then reversing from each corner to the center, or by moving around the plate to any part of its surface and reversing the test at each time, the multiplied error (if any) will be plainly shown by the test indicator, as the extension arm on the surface gage is in this case six times longer than the base of the gage. It is necessary that the test indicator should give exactly the same reading when moving forward, backward or sideways on any ordinary test. The indicator shown will stand this test without variation in the reading. It will be noticed in the illustration that a cutter is slipped on to the tail end of the extension arm for a counterbalance.

TESTING THE ACCURACY OF AN ANGLE PLATE

Fig. 179 illustrates a method of testing an angle plate. It will be seen that an adjustable angle plate is used for holding the plate to be tested; the adjustable plate, while preferable, is not, however, indispensable. After testing the angle plate in the position shown, and reversing it to bring the face up, the indicator attachment is reversed and the same

[1] Contributed to the *American Machinist* by J. H. Boulet.

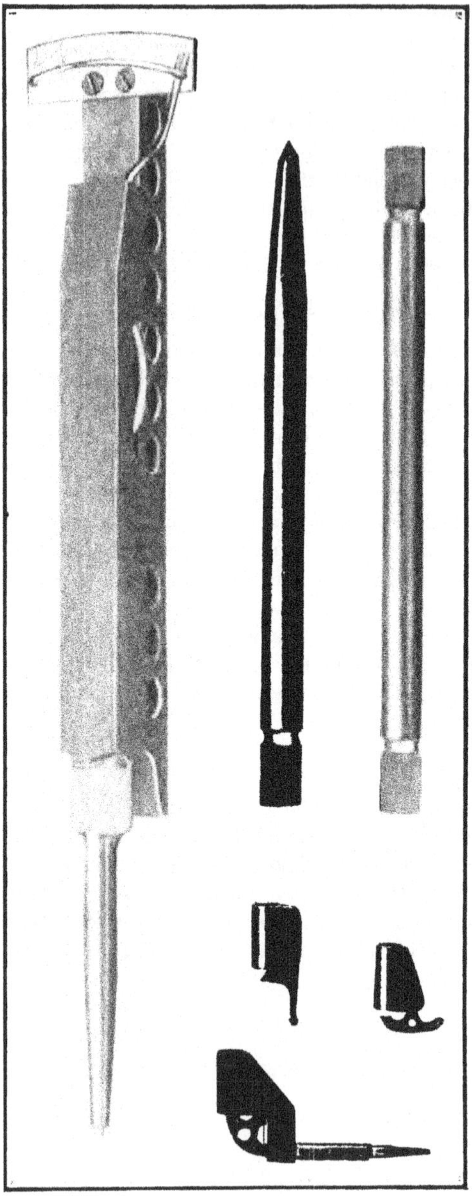

FIG. 177. — UNIVERSAL TEST INDICATOR

face tested in that position; if the two tests show the face to be correct, both angle plates are surely correct. On the contrary, if the angle plate being tested shows 0.002 inch high, for example, at the outer edge in the first test, and upon reversing the plate, face up, the same edge is found to be 0.002 inch low, the plate being tested is 0.002 inch out of square and the one used for clamping against is correct.

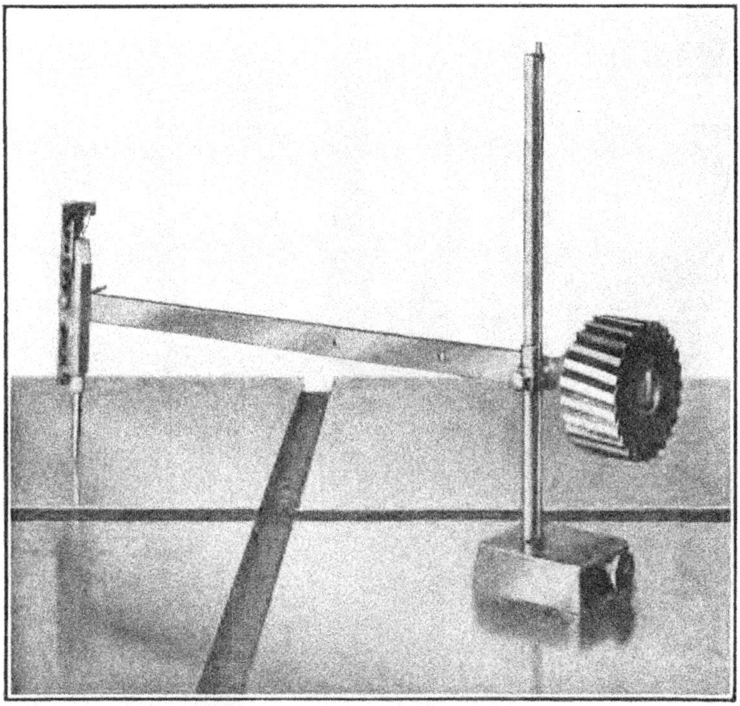

FIG. 178. — TESTING A SURFACE PLATE

Assume another test is made as shown in Fig. 179. Suppose the angle plate being inspected is apparently correct at the first test, but when reversed to bring the same face up, the second test shows 0.003 inch out; this will show that each of the angle plates is out of square 0.0015 inch, providing, of course, that they are of the same dimensions. If the adjustable plate, say, is larger than the one being tested, its total error will be proportionately greater.

Take another case which differs from the two above. Say in the first test (as shown in Fig. 179) the angle plate is 0.001 inch high at the outer edge, and when reversed face up for the second test it is found 0.004 inch low, at the same edge; that is, when in the first position, the error of one angle plate is subtracted from the error of the other plate, leaving a difference of 0.001 inch, while upon reversing the angle plate its error is added to the error of the adjustable plate, giving a total error of 0.004. Assuming that the two plates are of the same hight, the adjustable plate will therefore have an error of 0.0015 and the plate being tested an error of 0.0025, the sum of these two errors being 0.004 and the difference 0.001.

FIG. 179. — TESTING AN ANGLE PLATE

THREAD TESTS

Fig. 180 shows two different tests. At the left in the photograph is shown a way to test the lead of a threaded plug or tap on a surface plate with the aid of a hight gage and the universal test indicator. When placing the plug in the position shown, it is necessary to draw a center line on the threaded portion; if a tap is held instead, the edge of the flute can

FIG. 180. — TWO TESTS ON THREADED WORK

be used as a guide line. The test indicator is applied in the thread at one end of the plug or tap, transferring then to the hight gage, taking its reading and repeating the same operation at the other end of the plug or tap. By subtracting the lowest setting of the hight gage from the highest and counting the number of the threads which separate the two tests, the error in the lead, if there be any, will be easily determined.

The other test in Fig. 180 shows how to get the thread-caliper reading on a five-fluted or any other tap, without thread calipers. Place a piece of stubs steel wire in the thread; transfer the total hight as shown by the indicator to the hight gage; then subtract one and one-half times the size of the wire from the total hight found before, which gives the hight of the root of the V-thread. Then find the depth of the thread by dividing 0.866 inch by the number of threads per inch, and add one-half of this result to the hight of the root. Find the hight of the center of the tap by the shank and subtract this distance from the previous result obtained. Multiply the difference by two and the result is the caliper reading. Following is an example further to illustrate this test:

Size of wire = 0.063 inch.

Number of threads per inch = 10.

Hight of top of wire in V = 4.9998 inches.

Hight of center of tap = 4.0494 inches.

Constant for V-thread = 0.866 inch.

0.063 (diameter of wire) $\times 1\frac{1}{2}$ = 0.0945 inch.

4.9998 − 0.0945 = 4.9053 inches, hight of root of V above surface plate.

0.866 ÷ 10 = 0.0866 inch, depth of thread.

0.0866 inch ÷ 2 = 0.0433 inch.

4.9053 + 0.0433 = 4.9486, hight from surface plate to pitch line.

4.9486 inches − 4.0494 inches = 0.8992 inch.

0.8992 × 2 = 1.7984 inches, or the proper caliper reading.

1.7984 + 0.0866 inch = 1.885 inches, outside diameter of tap.

In Fig. 181 is shown another way of testing the lead of a thread plug or tap on a lathe by using the same test indicator with one of its regular attachments, which is placed in proper position, giving a direct-reading. By engaging the nut of the lead screw to the latter, after placing the instrument as shown in the illustration and revolving the lathe spindle as many times as desired, according to the length of the threads, any error as compared with the lead screw will be plainly shown.

THE INDICATOR IN THE MILLING MACHINE AND LATHE

Fig. 182 shows a method of testing manufactured parts on a fixture in which they have been milled, without the aid of any special gages.

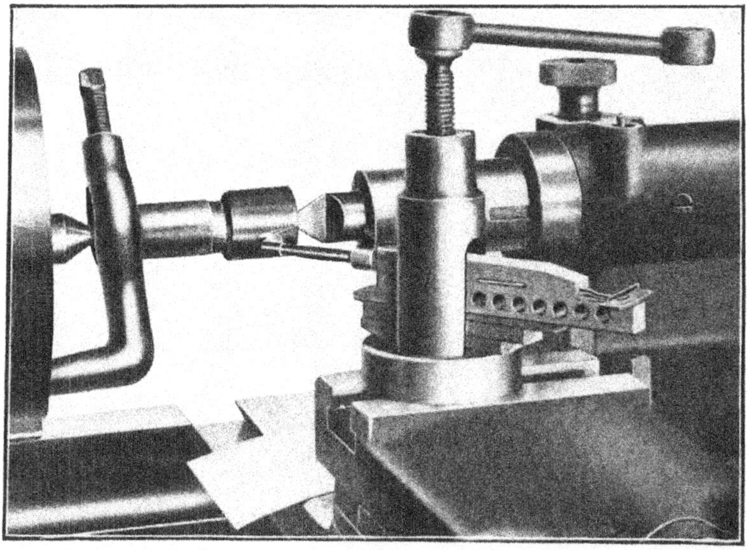

FIG. 181. — TESTING A THREADED PLUG IN THE LATHE

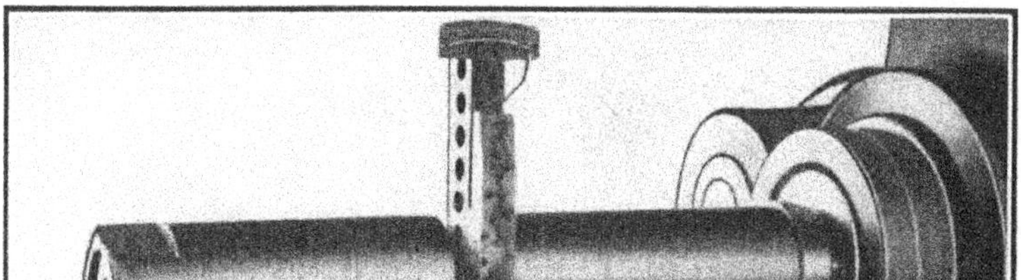

FIG. 182. — TESTING A JOB ON THE MILLER

[illegible]

[illegible] at one
[illegible] its read-
[illegible] of tap.
[illegible] and
[illegible] texts, the
[illegible]
[illegible] reading
[illegible] a piece
[illegible] by the
[illegible] the
[illegible] light
[illegible]
[illegible] this
[illegible] by
[illegible]
[illegible] Fol-
[illegible]

[illegible] surface

[illegible]

[illegible]

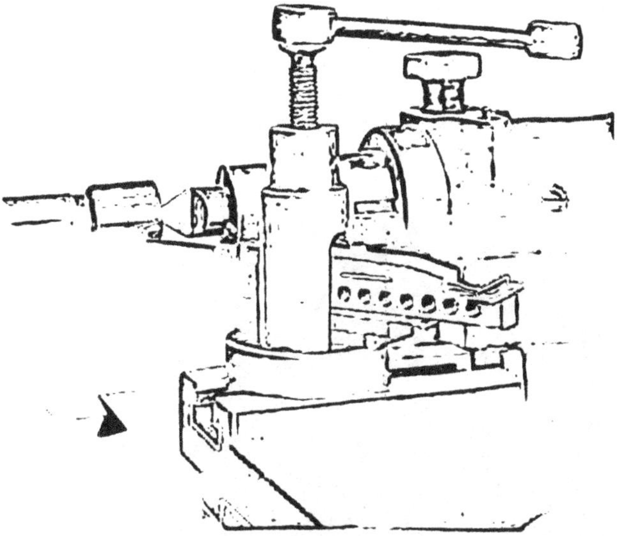

FIG. 181. — TESTING A THREADED PLUG IN THE LATHE

Fig. 183 shows how to apply the same instrument for setting a milling machine vise, or shoe. These two illustrations will suggest the variety of uses to which the instrument can be put, both in the manufacturing and tool-making departments. Fig. 184 shows how to test the truth of the inner face of collars, saws, cutters, etc., that can be reached through the hole or at the outer edge between the jaws of the chuck on the lathe or grinding machine.

FIG. 183. — SETTING A MILLER VISE

FINDING THE SIZES OF HOLES IN JIGS

In Fig. 185 is shown a method of obtaining the sizes of holes without the aid of plugs while inspecting a jig, the tests being made by reaching at the inside (near the steel seat shown) for the most accurate results. It is very often the case that a jig hole has to be bored to an odd size, and when the toolmaker and inspector run across such a case (before the bushings are to be put in) it is quite common practice to go to the expense of making special plugs. By taking the hight from the inside at the lowest part of the bored hole or lining bushing, as shown, and transferring to the vernier hight gage, then taking the hight at the upper part of the hole by eversing the regular attachment on the indicator, and subtracting one hight from the other, the size of the hole is obtained. Adding one-half of the diameter of the hole to the lower hight already obtained gives the center hight.

OTHER USES IN JIG WORK

The illustration in Fig. 186 shows how to obtain a measurement in a box jig, as in taking the hight of a steel collar or seat around the stud, where the work is to be located. The test indicator is mounted on a surface gage with an extension so as to hold the instrument in an upright position, and the hight when taken is transferred to the hight gage, giving the starting point of all measurements to be taken on the jig.

FIG. 184. — INDICATING REAR FACE OF A CHUCK JOB

Fig. 187 shows another way of reaching into a box jig (placed in an upright position) for measurements to be taken and transferred to the hight gage. In testing drill jigs, the most reliable results are obtained by working from the inside of the jig as much as possible. If a hight has to be taken at the outside with a test bar or plug inserted in holes in the jig, great care should be taken to have the bar perfectly parallel with the surface plate, as very often the holes are bored to a slight angle due to springing the jig when clamping it.

LOCATING A JIG FOR BORING

In general tool work it is often the case that a drill jig laid out with care is plenty good enough to work to, that is, if the center-punch marks are properly located in the lay-out; the next thing is to follow these center-punch marks. The center marks can be located for setting the work, first roughly with the center in the spindle of the machine, then by placing the pointed test bar (shown under the indicator in Fig. 177) in the center-

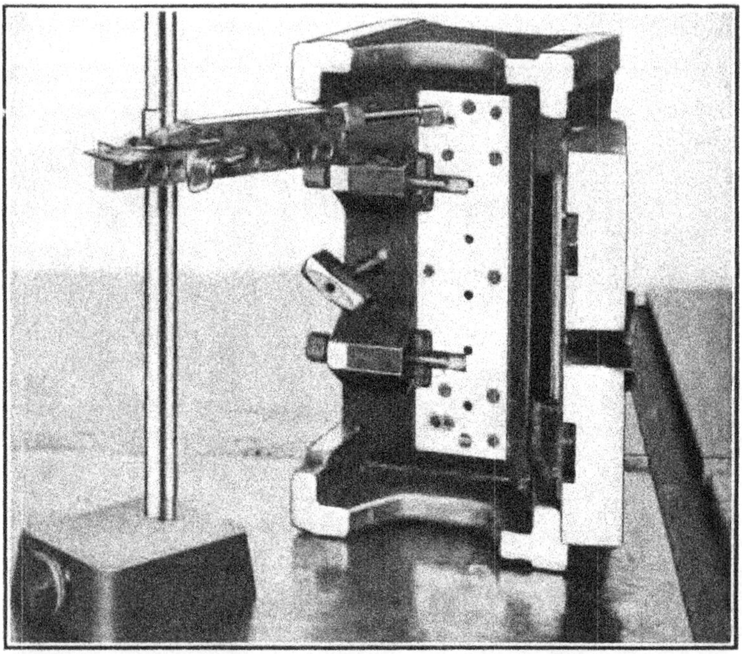

FIG. 185. — GETTING SIZE AND CENTER HIGHT OF A HOLE IN A JIG

punch mark in the jig with the outer end of the bar on the boring-machine spindle center, the latter being perfectly true. Of course it is necessary to secure close alinement of the test bar both ways. To obtain this alinement, the test indicator may be clamped on the end of the spindle of the machine and brought into contact with the test bar as near to the work as possible, the machine spindle being turned by hand and the readings of the indicator noted at the four quarters; this will plainly show the position of the jig and what adjustments are to be made to locate it properly.

Another way to locate by the center-punch mark is to aline the test

FIG. 187

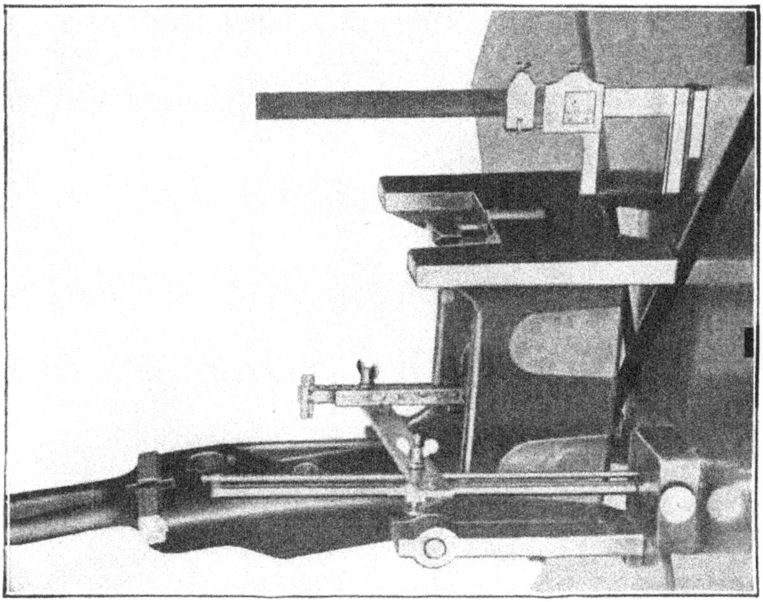

FIG. 186

OTHER APPLICATIONS IN CONNECTION WITH JIG WORK

bar in position (from the carriage of the machine) with the indicator mounted on the surface gage, using one of the regular attachments to facilitate setting the test bar parallel with the plate. For side alinement an angle plate may be placed on the carriage parallel with the spindle, using the upright surface of the angle plate with the test indicator mounted on the surface gage as before. Thus the test bar is readily set parallel with the surface of the angle plate, giving the proper side location.

The button method of locating jigs is of course one of the most reliable used. For setting the jig correctly on the machine carriage the test indicator may be clamped on the end of the spindle, reaching the button by means of one of the indicator attachments; then upon revolving the indicator or testing at the four quarters of the circle, the readings will be alike when the proper location of the jig is obtained.

These methods are applicable to milling-machine as well as boring-mill work.

CHAPTER XVII

A Universal Indicator and some of Its Applications

In addition to the information given in Chapter XVI on the uses of the test indicator, some illustrations are here presented showing a few applications of another universal indicator which should be of interest to toolmakers and others. This instrument is manufactured by Henry Koch & Son and its principal features are revealed by Figs. 188 and 189.

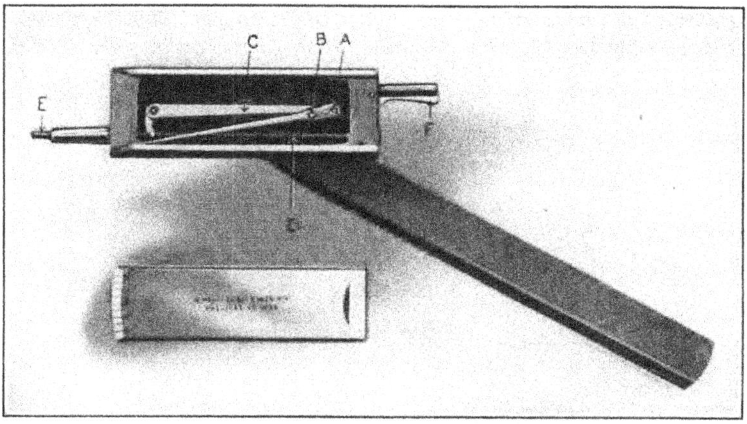

FIG. 188. — TEST INDICATOR WITH COVER PLATE REMOVED

The indicator proper consists of a rectangular steel case which carries the pointer, lever, and other mechanism and is adapted to swivel freely at the end of a holder which may be used in the lathe, shaper, or miller. The indicator head may also be attached to the post of a surface gage and used for a variety of bench-plate work, or it may be placed on either of its flat sides and operated in conjunction with size blocks and parallels for testing work. The contact point for external and internal operations are at opposite ends of the indicator, and by loosening the clamp nut underneath the device is readily swiveled end for end to bring either point desired into contact with the work.

ADJUSTMENT ON HOLDER

It will be seen that the indicator may be swung around and clamped at any desired angle on the holder or shank, and, as there is a hole drilled crosswise through the latter at the rear end, the device may be attached at the side of the shank and then swiveled to any position required. This feature is of advantage especially when the instrument is to be clamped between collars on a milling-machine arbor, as the holder may then be

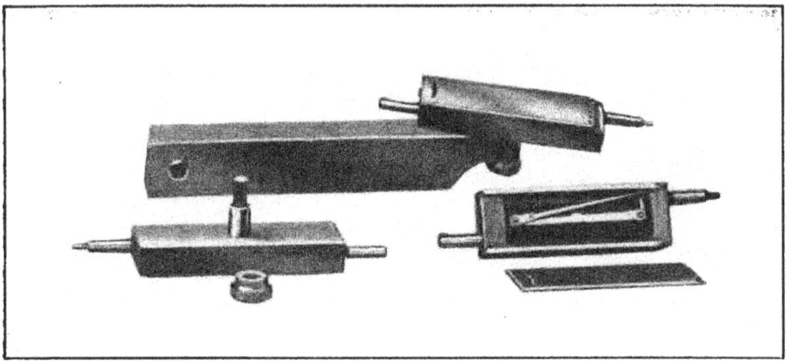

FIG. 189. — TOP AND BOTTOM VIEWS OF TEST INDICATOR

placed with its broad, flat side against the arbor and with the narrower edges gripped securely between the collars at either side. The indicator thus constitutes a universal device whose numerous applications are but briefly outlined in this article. The illustrations, Figs. 190 to 194, showing some of its uses will, of course, suggest many other ways in which it can be employed to advantage by the toolmaker.

THE MECHANISM

The indicator mechanism, which is simple, light and well protected by the rectangular case and its sliding cover (on the beveled end of which are graduations reading to thousandths) may be briefly described as follows: The indicating pointer is pivoted at *A*, Fig. 188, and has a pin at *B*, which contacts with the edge of the long arm of rocker lever *C*. Against pin *B* rests the end of a spring coiled about the post under *A* and tending to swing the outer end of the indicating pointer across the opening in which it plays when in operation, or to a position directly opposite that in which it is actually shown in the engraving. When the instrument is at rest, however, the pointer is maintained in the position shown by plunger *D*, which is pressed forward by the compression spring shown, and a collar on which contacts with the short arm of lever *C*.

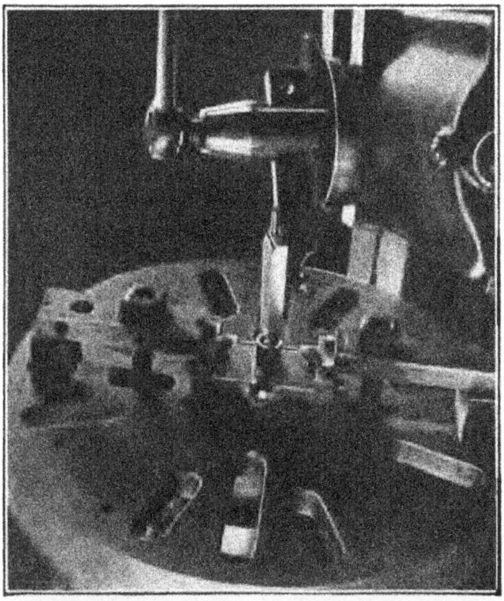

FIG. 191

FIG. 190

THE INDICATOR IN THE LATHE

[illegible]

FIG. 193

FIG. 192

THE INDICATOR IN THE MILLER

CONTROL OF POINTER

The outer end of the plunger forms the contact point *E* for external work; and it is obvious that when it is pressed back by the action of the piece which is being tested against the point, the indicator pointer will be operated by the light spring under *A* to show any errors in the work. That is, the pointer is actuated in one direction by the light spring referred to, but the spring can operate the pointer only when the contact plunger *D* and its heavier spring are pressed back by the work being tested. Thus the pointer is always controlled between the two springs, giving it a quick, sensitive action, and at the same time the contact plunger *D* can be pressed inward to its limit without increasing the constant, delicate pressure applied to the pointer by the operating spring coiled about pin *A*. There is, therefore, no possibility of the instrument being injured if brought against the work too abruptly, or with too great pressure.

FIG. 194. — MAKING A MEASUREMENT IN A JIG

APPLICATION TO INTERNAL WORK

Internal indicating, as well as some external testing operations, are accomplished by the contact piece *F* which is in the form of a bell crank with the concealed end abutting against the inner face of a collar at the rear end of plunger *D*. When this end of the instrument is in use the action of the mechanism is precisely the same as when the end *E* is employed. In one case the spring plunger *D* is drawn back from the rear; in the other it is pushed back from the front.

LATHE OPERATIONS

In Figs. 190 and 191 are shown two applications of this indicator to lathe work which has to be trued up prior to machining. The first illustration represents a method of centering a pilot in a punch which is held in a special face-plate fixture.

The illustration in Fig. 191 shows a method of locating a jig on the face plate prior to boring a bushing hole; the indicator in this case is brought into contact with a button which is located at the desired point on the jig leaf.

THE INDICATOR AS USED ON THE MILLER

Fig. 192 illustrates the manner in which the indicator may be clamped to the milling-machine arbor while testing a vise jaw to determine if the vise is properly set. Fig. 193 shows the indicator held in a chuck on the miller spindle while a jig on the table is being located by means of a button on which the indicator contact point rides.

Fig. 194 illustrates a surface-plate job where it is desired to transfer a measurement from a hight gage to a surface inside a jig which it is difficult to reach by ordinary means. The indicator in this instance is attached to a surface gage and extends down inside the jig in a perpendicular position, as illustrated.

CHAPTER XVIII

A New Swedish Combination Gaging System[1]

A REMARKABLE set of gages manufactured by C. E. Johansson, Eskilstuna, Sweden, and sent to the *American Machinist* office for inspection by the Grönkvist Drill Chuck Company, are illustrated in the accompanying engravings.

Fig. 195 shows the case containing the 81 gages. It measures $10 \times 15\frac{3}{4}$ inches and weighs $7\frac{1}{2}$ pounds. The first series of gages, 9 in number, are from 0.1001- to 0.1009-inch in thickness, rising by 0.0001. The second series, 49 in number, are from 0.101- to 0.149-inch in thickness, rising by 0.001. The third series, 20 in number, are from 0.050- to 1.000-inch in thickness, rising by 0.050. The fourth series, 3 in number, from 2 to 4 inches, rising by 1 inch.

HOW THEY ARE USED

These gages are used either singly or in multiple. With them the maker claims that over 80,000 different sizes of gages can be made, and, in all probability, this is rather a low estimate. The object of the set is to give a practically unlimited series of sizes rising by increments of 0.0001-from 0.1001-inch to about 26 inches. Some of the things done with them and shown by the half-tones would seem to indicate that the error must be very small. The photographs were made in the presence of one of the editors of the *American Machinist.*

THE ACCURATE FLAT SURFACES

The surfaces of these gages are so flat and smooth that if two or more gages be wrung together so as to expel the air they will stick together. Fig. 196 shows 36 gages thus wrung together and held horizontally. Fig. 197 shows 21 pieces wrung together and held vertically. The faces in contact measure about $\frac{3}{8} \times 1\frac{3}{8}$ inch. Fig. 198 shows 22 pieces held with the narrow way vertical. Fig. 199 shows one large piece and two small pieces aggregating the same measurements. On the outside faces two more pieces are wrung on, lapping half on the single piece and half on the built-up piece. The whole may be supported by any one of the pieces without falling apart. Two other views of this piece are shown in Fig. 200. An indefinite number of combinations may be successfully subjected

[1] E. A. Suverkrop in the *American Machinist.*

FIG. 195. — THE GAGES IN THEIR BOX

to this test, which, as regards the accuracy of the pieces, is, of course, almost absolute.

USED AS PLUG GAGES

Fig. 201 shows the way these gages are used in multiple for plug gages. The two outside pieces are lapped to a circle. The two wrung together will measure a half-inch hole. Any of the flat gages can be used between the outside pieces to make a gage for any size of hole desired. Fig. 202

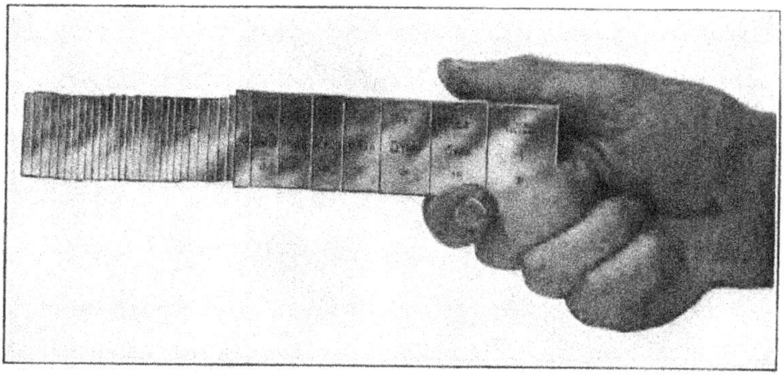

FIG. 196. — THIRTY-SIX GAGES WRUNG TOGETHER AND HELD HORIZONTALLY

shows a 1-inch snap gage with a built-up inch in it — 0.5 + 0.2 + 0.050 + 0.150 + 0.100. The peculiarity about this is that the *feel* of the built-up gage in the snap gage is just the same as the *feel* of the single 1-inch piece, and this holds for any combination aggregating one inch, the number of pieces not affecting the *feel* in the slightest. Any mechanic will understand instinctively the meaning of this statement and its testimony to the accuracy of the gages.

The feel of the pieces when wrung together is like that of strongly magnetized steel, and until one has played with these gages for quite a while it is impossible to eliminate this impression of magnetism. That they are not magnetic is shown by the fact that they show no signs of adhering unless wrung together.

WHAT HOLDS THE GAGES TOGETHER

While handling the gages one is surprised at the force necessary to separate two of them when pulling at right angles to the contacting surfaces. This has caused some argument as to whether the adhesion was due to atmospheric pressure or molecular attraction, or both. To test this two small toolmaker's clamps and a 12-pound spring balance were borrowed. The pressure on the area of the contacting surfaces, allowing 15 pounds air pressure per square inch, worked out to 7.6 pounds. It was found that the 12-pound spring balance was not powerful enough to register the force necessary to pull the pieces apart, although a strain of $11\frac{3}{4}$ pounds was put upon the pieces for a period of about 20 minutes.

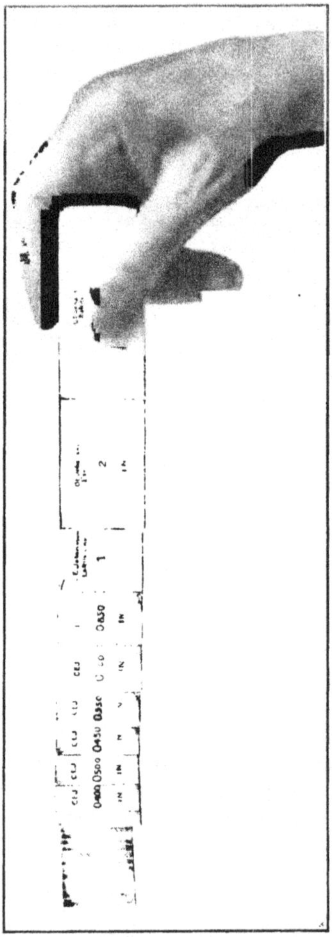

FIG. 197. — TWENTY-ONE PIECES WRUNG TOGETHER AND SUSPENDED VERTICALLY

Subsequently a larger balance was used and the surfaces held together, on two occasions, until a force of 16 pounds, or over 31 pounds per square inch, was reached. This would prove conclusively that something more than the mere atmospheric pressure holds the pieces together.

HOW THE TEST WAS MADE

The test was crude, there was no way of getting the clamps exactly in the centers of the pieces, or the pull straight. The spring balance was hung from above, as shown in Fig. 203. The clamps were tightened on

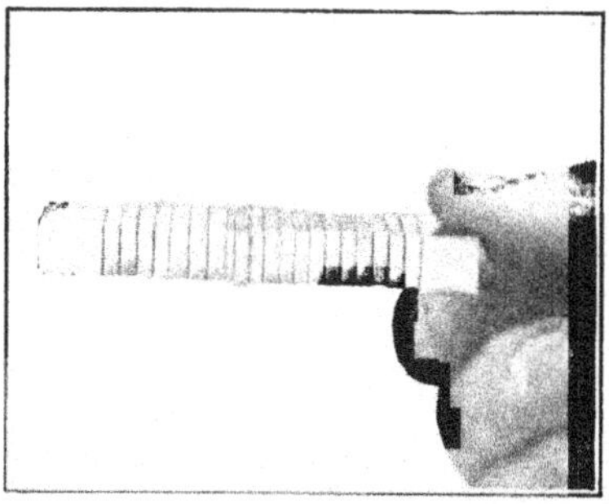

FIG. 198. — TWENTY-TWO GAGES WRUNG TOGETHER AND HELD WITH THE NARROW WAY VERTICAL

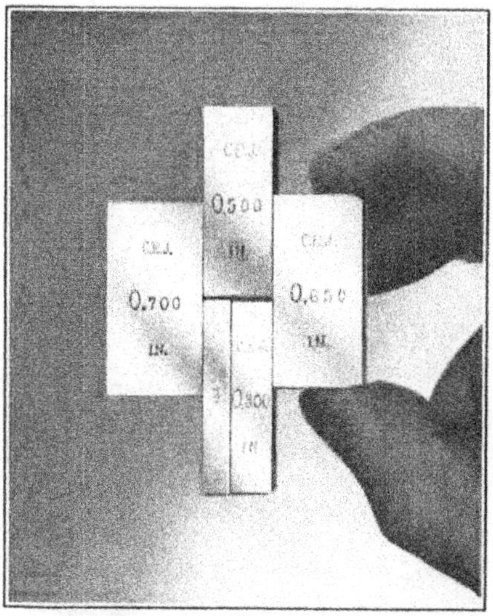

FIG. 199. — FIVE PIECES WRUNG TOGETHER

the gages with a piece of chamois leather betweeen the jaws and the gages. A small screw eye was screwed into a board below the balance. One piece of string, leading to the left, was tied to the lower clamp as a

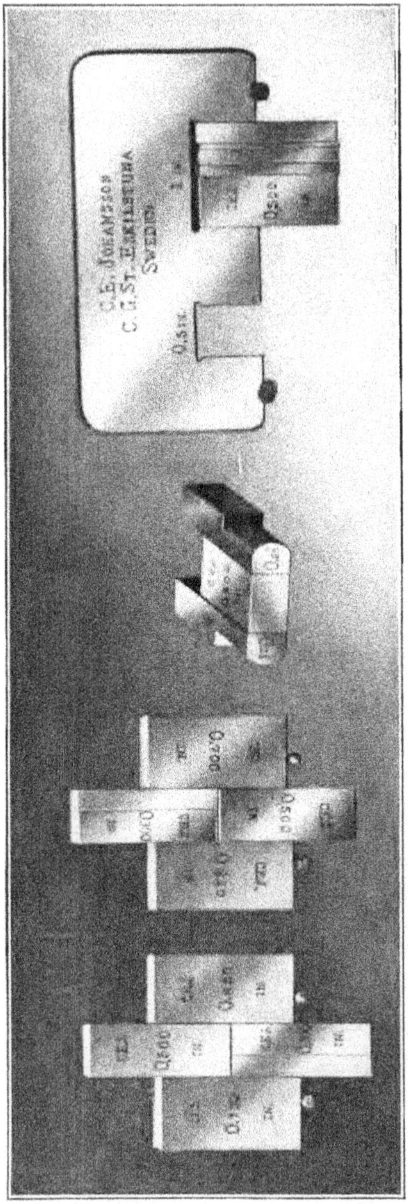

FIG. 200. — TWO OTHER VIEWS OF THE FIVE PIECES WRUNG TOGETHER AND SUPPORTED ON TWO NAILS

FIG. 201. — ONE-INCH BUILT-UP GAGE

FIG. 202. — SNAP GAGE WITH BUILT-UP INCH

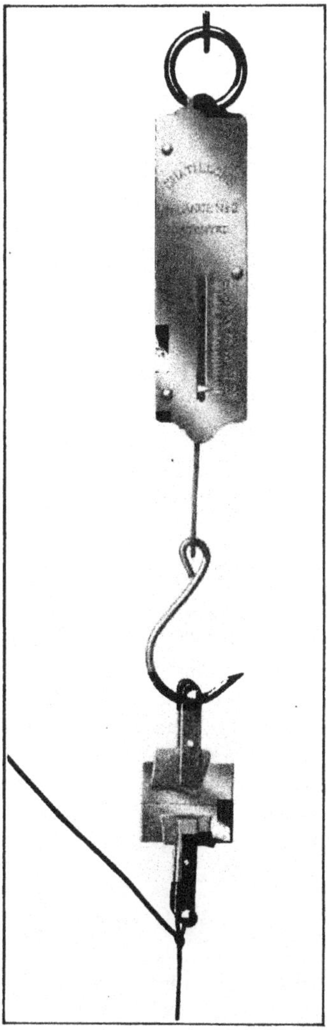

FIG. 203.—TWO GAGES WRUNG TOGETHER AND TAKING A STRAIN OF $11\frac{3}{4}$ POUNDS OR 22 POUNDS PER SQUARE INCH

safety device so that the gage would be safe from striking anything when it was pulled away. The other string was tied to the lower clamp, and led down through the screw eye. The illustration shows that in this case there was a strain of $11\frac{3}{4}$ pounds on the string, which is equivalent to over 22 pounds per square inch.

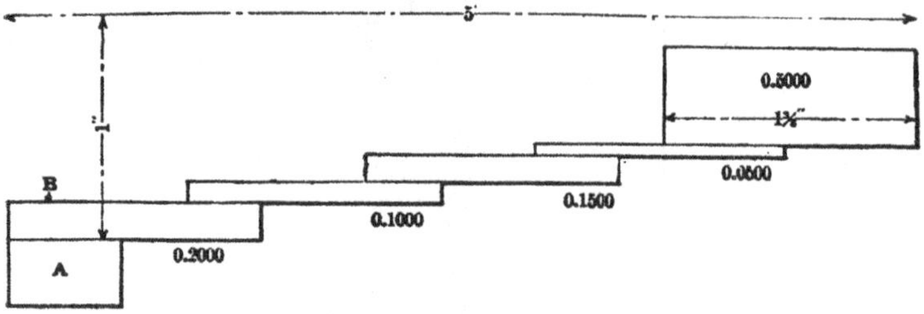

FIG. 204. — OFFSET GAGES SUPPORTED FROM ONE END

After the photographs were made other things that could be done with the gages were discovered. One of these is shown at Fig. 204. The pile of gages was supported at *A*, with sufficient weight at *B* to hold the pieces horizontally.

HOW THE GAGES ARE SELECTED FOR A GIVEN SIZE

When picking out gages for a desired size it is best to follow a simple system. For instance, take the size 3.1416 inches. The best way to begin with the selection of gages is to get rid of the fraction first, thus we select 0.1006 as our first component and, subtracting 0.1006 from 3.1416 we get 3.0410. Next we get rid of that 0.0410, but, as our smallest gage is 0.0500 we must take one larger than 0.0410, subtracting the difference from the size to the left of the decimal place, which in this instance is 3 inches. We select gage 0.1410 and subtract it from our last remainder 3.0410, which gives 2.9000. The next two gages 2.000 and 0.9000 are selected without trouble; thus the operation is:

$$
\begin{array}{ll}
3.1416 & \\
\underline{0.1006} & = \text{1st increment.} \\
3.0410 & \\
\underline{0.1410} & = \text{2d increment.} \\
2.9000 & \\
\underline{0.9000} & = \text{3d increment.} \\
2.0000 & = \text{4th increment.}
\end{array}
$$

THE SAME SIZE AVAILABLE BY USING OTHER INCREMENTS

At the same time that this set aggregating 3.1416 is in use by one man, it is just possible that another man in the shop might want the same measurement. In that case another gage can be made up using other increments, as shown in the table:

0.1001
0.1005
0.1250
0.1260
0.1440
0.1460
0.4000
0.7000
0.3000
1.0000
3.1416

It is often very handy to have several gages of the same size available at the same time. The case cited, 3.1416, of course is unusual, but take for instance the size 0.60. This may be made up as follows: 0.550 + 0.050; 0.450 + 0.150; 0.400 + 0.200; 0.350 + 0.250; 0.500 + 0.100, etc.

BINARY FRACTIONS OBTAINABLE

Regular shop sizes rising by eighths, sixteenths, thirty-seconds, sixty-fourths, etc., and plus or minus one or more thousandths, are easily obtainabie. Take for instance $1\frac{5}{8}$ inches. Gages 1 inch, 0.500 and 0.125 will give it. Suppose instead of $1\frac{5}{8}$ inches we want 0.002 larger or 1.627. We substitute 0.127 for the 0.125 gage of the former example and this gives us what we are looking for. The same size can be obtained by several other combinations. Suppose we wish 0.002 less than $1\frac{5}{8}$ inches or 1.623. Gages 1.000 + 0.500 + 0.123 would give us what we are looking for, and in a similar way any size may be obtained rising by thousandths or tenths of thousandths. Repeated trials have failed to find a size within the limits of the set that cannot be obtained, and usually by several combinations.

It is right here that we have the essential difference between these and other fixed gages. The usual pattern of fixed gages gives a defined set of sizes which does not include the plus allowance for a forced fit nor the minus allowance for a running fit, whereas we have here a means of obtaining any size required, including any allowance for forcing or running fits, down to the smallest increment of the set. Of course this use of the pieces in multiple requires a high degree of accuracy if the results are to be satisfactory, but the tests already described certainly indicate that the requirements in this respect have been met.

SOME OF THE USES THEY CAN BE PUT TO

These gages can be used for hight gages, distance blocks, for testing work, making working gages, etc., etc. Figs. 205 to 211 show a few of the many uses they can be put to.

Fig. 205 shows a temporary snap gage. The two inner pieces, 0.1001 and 0.3500, are wrung together and then the two outer "jaws" are wrung

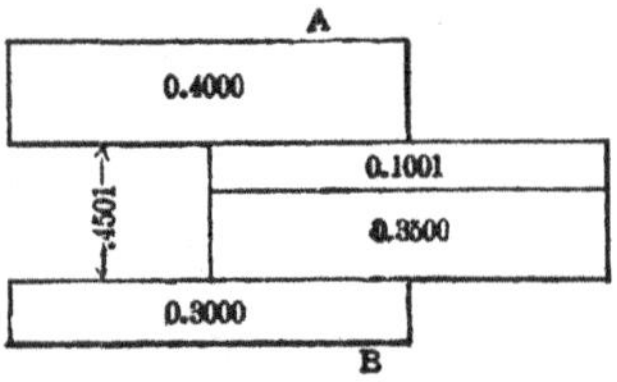

FIG. 205. — TEMPORARY SNAP GAGE

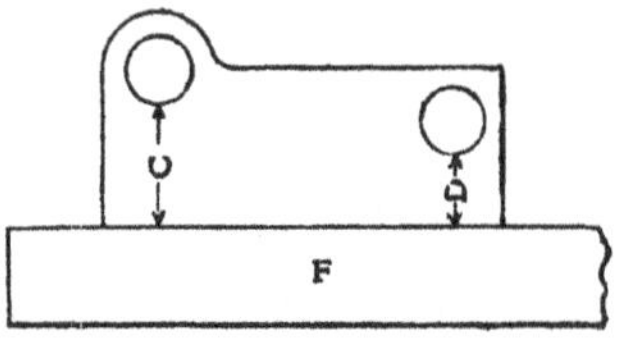

FIG. 206. — MEASURING FROM SURFACE PLATE

on. If desired any ordinary toolmaker's clamp may be used at *A B* to bind the pieces together. It is, however, hardly necessary as they hold together very firmly without any clamping. Fig. 206 shows a piece with two holes in it to be tested. Plugs are put in the two holes and the piece is put on a surface plate. The distances *C* and *D* are easily measured by slipping the gages between the plugs and the surface plate. Fig. 207

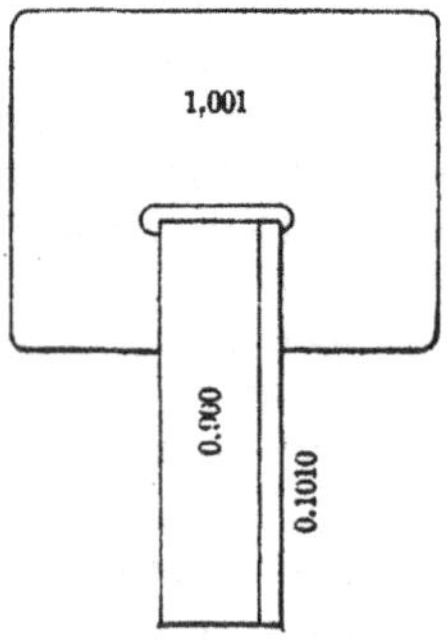

FIG. 207. — TESTING A SNAP GAGE

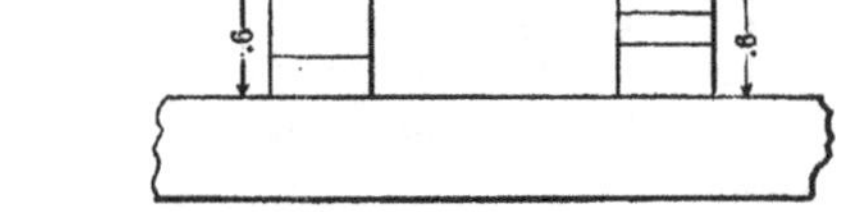

FIG. 208. — LAYING OUT WORK

shows the testing of a 1.001 snap gage. Fig. 208 shows a piece on which two lines are to be scribed 0.600 inch apart. The piece is put on the surface plate and the first line *G* is scribed with the scriber *I*. The piece is then blocked up on the gages and the second line H is scribed. Fig. 209 shows a snap gage having two openings which bear a fixed relation to each other. Fig. 210 shows practically the same thing except that the gages are applied direct to the work, which in this case is a shaft with three rings on it.

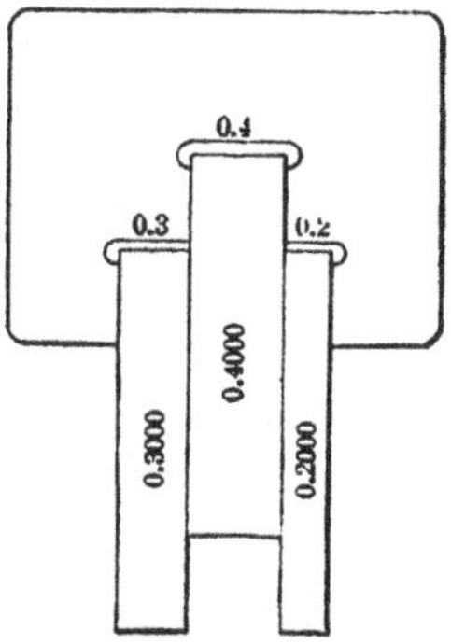

FIG. 209. — TESTING DOUBLE SNAP GAGE

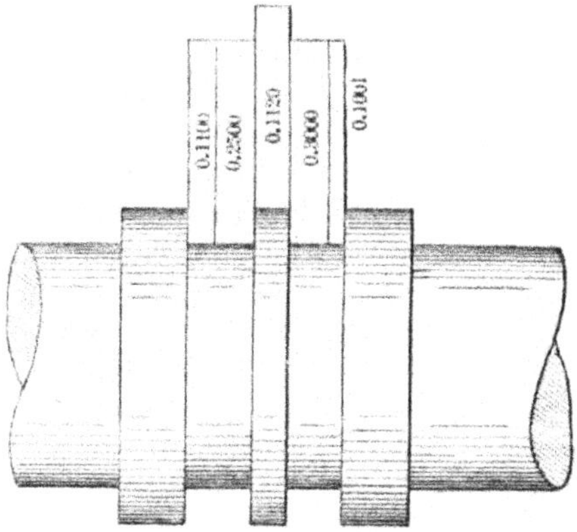

FIG. 210. — TESTING SHAFT COLLARS

Fig. 211 shows the gages used on a face plate in conjunction with two parallels to locate the work, so that two holes may be bored in it in accurate relationship to one another.

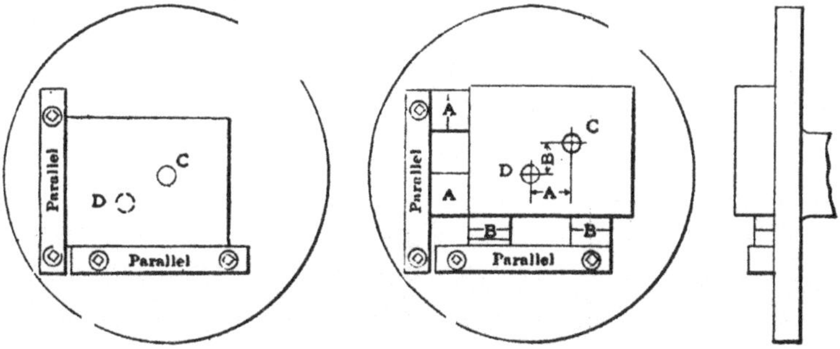

FIG. 211. — LOCATING WORK ON FACE PLATE

THE GAGES ARE SELF-CHECKING

One of the most valuable features of this set of gages is the fact that they are self-checking. If a built-up gage does not agree with a single gage, then the increment which is incorrect can easily be discovered by a process of elimination. For instance, suppose we find that 0.600 + 0.400

inch does not fit snugly in the inch snap gage made up as in Fig. 205 but with the 1-inch piece between the jaws. We could then substitute two gages for say the 0.600. If the gage thus made up fitted all right it would show that the 0.600 gage was under size. If the *feel* remained the same, then obviously the 0.400 gage would be small. Substituting two gages, say 0.250 and 0.150 inch for the 0.400, would, if they fitted the snap gage, prove that the 0.400 was small.

It will be seen that we have here not only a new style of gage but a new system of gaging and of originating shop sizes. The use of gages in multiple is, of course, common enough, but not on the lines here laid down, which, so far as known, are new and give sizes which heretofore it has only been possible to get from measuring machines or micrometers. The smallest increment is small enough for determination of not only limits but tolerances, and the gages would seem to furnish all needed means for determining working gages of any required degree of accuracy.

CHAPTER XIX

SETTING, LAYING OUT AND TESTING WORK WITH THE SWEDISH GAGES[1]

THOSE who have read the preceding chapter on the Johansson system of gages cannot fail to be interested in the additional illustrations presented here showing some further applications of these gages. As stated in Chapter XVIII, the basic idea of the system is the making of gages of such sizes that when used in combinations any dimension whatever, down to the smallest limit of the set, — one ten-thousandth of an inch, — may be obtained. This being the case, it is obvious that the gages may be used not only for originating sizes, but also for testing work already done, down to the degree of accuracy represented by one ten-thousandth of an inch.

As previously shown, the gages are essentially internal gages, but the appliances shown herewith enable them to be used as either internal or external gages, directly, for the setting of calipers, or, as in Fig. 212, for the adjustment of a supplementary limit gage for the testing of work.

LIMIT GAGES

The gage shown in Fig. 212 consists of the lower jaw projecting on each end, the sides which connect this to the top piece, the sliding jaw and the adjusting screws.

As shown in Fig. 212, the gage is being set to a maximum of 0.550 inch and a minimum of 0.545, a limit of 0.005 inch. The maximum is obtained by a single block and the minimum by using 0.400 and 0.145 in combination. The sliding jaw is moved down after loosening the nurled nut, when it is moved into any desired contact by the long screws. These are then clamped in position by the small screws shown in the upper cross piece. The minimum jaw is moved into contact with its block, after loosening the two screws, and then locked in a similar manner. When set it becomes a rigid limit gage of the desired size as, in addition to the long screws, there is a hexagon milled nut behind the nurled head of the screw; the sliding jaws can be clamped solidly by a wrench. The other end of this wrench is a screwdriver for tightening the adjusting screws. It requires little imagination to think of many cases where this would give just the

[1] Fred H. Colvin in the *American Machinist.*

gage wanted, varying the limit at will, yet where it would be impossible to stop to make a limit gage in the regular way or to pay for all the odd sizes one might need.

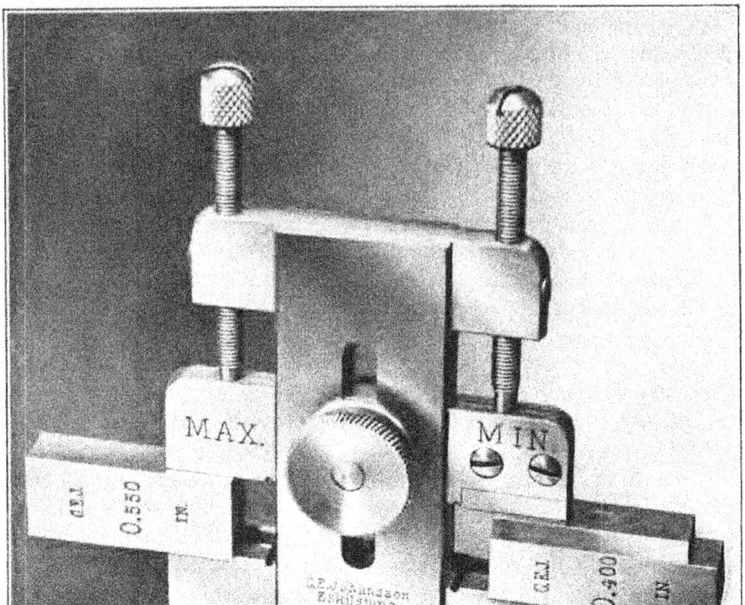

FIG. 212. — THE UNIVERSAL LIMIT GAGE

STANDARD GAGES IN LAYING OFF WORK

Fig. 213 shows a familiar use of a standard block or distance piece under the shaper tool, and other gages, singly and in combination, are used to locate the other depths. The widths can also be measured equally well, as will be seen later. Fig. 214 shows the same piece being tested after removal from the chuck by laying a straight-edge across the top.

INSIDE AND OUTSIDE CALIPER GAGES

For converting these gages into outside or inside caliper gages without number, the clamp or holder shown in Figs. 215 and 216 is provided. This describes itself except for the fact that the screw runs through a split nut so that by pressing on the checkered ends, the nut opens, and the screw is slipped to very nearly the desired position, when releasing the ends allows the nut to engage the screw and the last adjustments to be made.

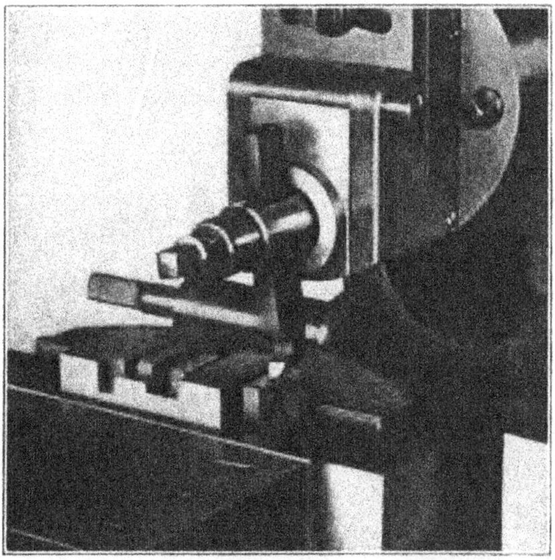

FIG. 213. — TESTING WORK IN THE SHAPER

As shown, the two projecting blocks are 60.5 millimeters apart, as these happen to be metric blocks. The other gages shown are, however, made to English sizes.

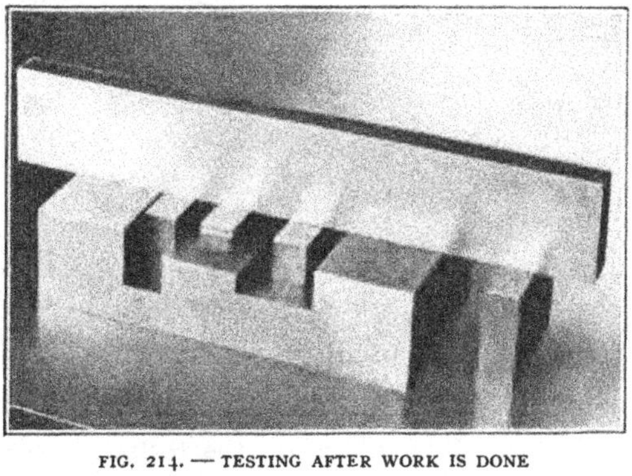

FIG. 214. — TESTING AFTER WORK IS DONE

Fig. 217 shows the use of this holder for setting calipers from the standard gages. The outside calipers at the left and the inside calipers at the right are both set to the same size, 3.25 inches, one using the outside of the two blocks and the other the inside of two other blocks clamped at

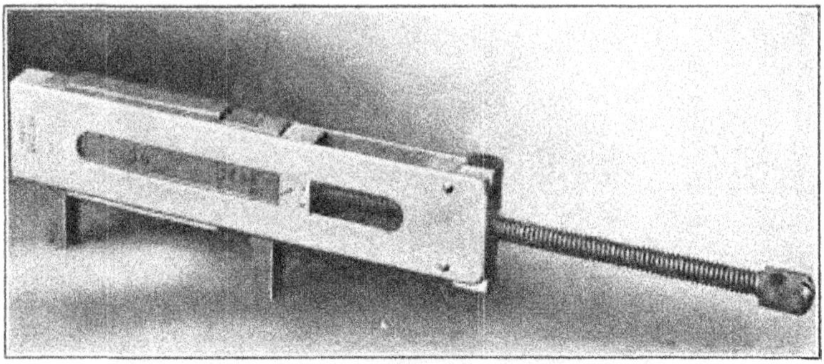

FIG. 215. — HOLDER FOR BUILDING UP GAGES

the ends of the distance pieces. These outside blocks have a radius equal to their thickness, so that when two of the same size are put together they form a plug gage of that size. Used in connection with other blocks they give a line contact inside a large hole or ring, as shown in Fig. 216.

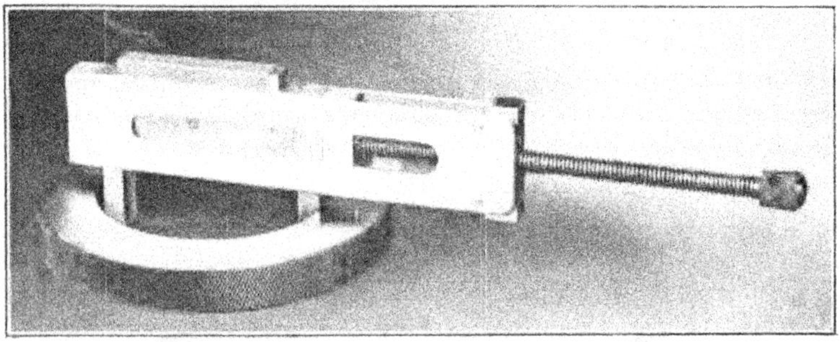

FIG. 216. — HOLDER AND ITS USE IN GAGING RINGS

Another application is shown in Fig. 218, where four blocks are built up to make 7.985 inches between the measuring or end pieces. As a means of setting end-measuring gages or of making and testing solid end or wire gages, this is extremely useful on account of its wide range.

The remaining figures show other uses for a set of standard gages such as these, and they are all interesting to the toolmaker who must work to close measurements and have ways and means for testing the results obtained.

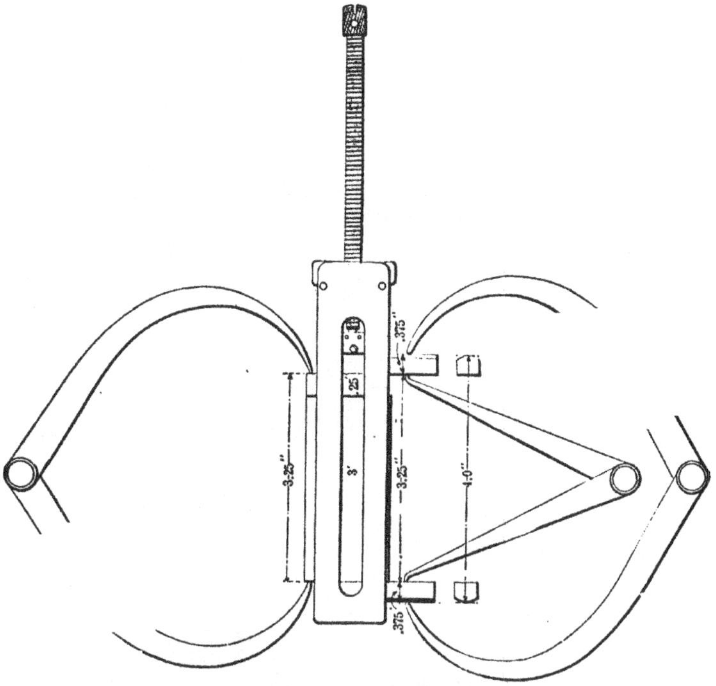

FIG. 217. — USING GAGES TO SET CALIPERS

In Fig. 219 is shown a method of laying off lines and distances by raising the piece of work on the proper combination of gages to secure the right hight. The surface-gage pointer is set for the highest line when the piece of work is resting on the surface plate, and other lines are located by putting known gages under the work. The many combinations allow

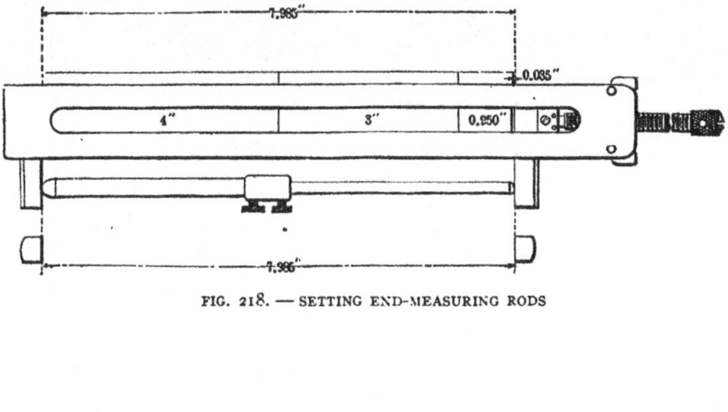

FIG. 218. — SETTING END-MEASURING RODS

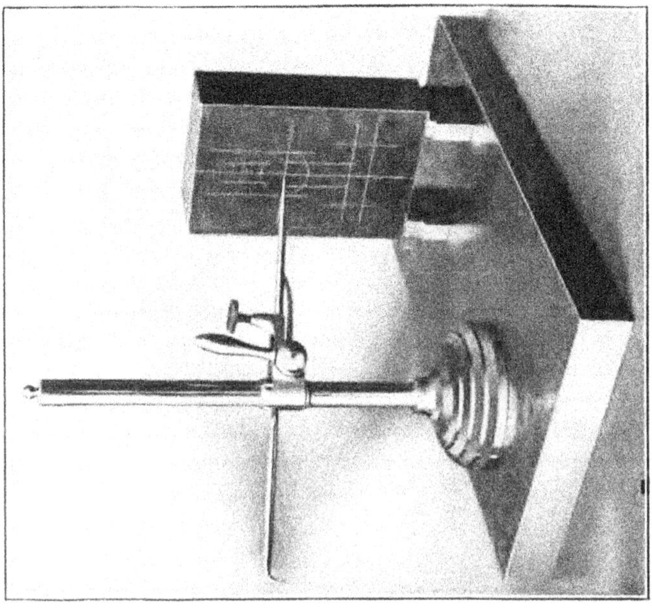

FIG. 219. — LOCATING POINTS WITH SURFACE GAGE

FIG. 220. — TESTING ACCURACY OF A DIE

any distances to be obtained, and there is a certainty of exactness in results that is not often obtained by ordinary methods. Incidentally this shows a form of Swedish surface gage which is a little different from the ordinary.

TESTING THE ACCURACY OF WORK

In Fig. 220 the accuracy of the work which was laid out with the surface gage is being tested by various combinations of gages. Every part can be measured accurately and with relation to the rest of the piece, as indicated by the upper portion, where the diameter of the hole is found by adding the thickness of the two plugs and the flat gage blocks; at the same time it shows the exact relation of the slot to the hole. Other combinations below give the measurement of the various openings.

FIG. 221. — TESTING SURFACES AND DIMENSIONS

A great variety of measurements are being tested in Fig. 221, which is some sort of a die block or gage that has been finished by milling. The distance of the two plugs or mandrels from the surface plate is measured by the blocks in front, while the pieces top and bottom of the plugs show whether they are in the desired location with reference to the top and bottom of the slot. The width of the slot is measured by the blocks at the left. The distance from the top of the large plug to the upper cross slot is also measured, and in the same way the slot itself could be checked up. Using the holder as shown in Figs. 215, 216, 217 and 218, the outside of this piece of work could be measured in the same way.

INDEX

A

B

C

D

Spotted something that isn't quite right?

Let us know and we'll fix it for future reprints.
And as a thank you for your diligence, we'll
send you a free Sothis book of your choice.

publisher@sothis.press

This page intentionally left blank

Made in the USA
Monee, IL
08 November 2026

68001564R00142